CAMBRIDGE LIBRARY COLLECTION

Books of enduring scholarly value

Cambridge

The city of Cambridge received its royal charter in 1201, having already been home to Britons, Romans and Anglo-Saxons for many centuries. Cambridge University was founded soon afterwards and celebrates its octocentenary in 2009. This series explores the history and influence of Cambridge as a centre of science, learning, and discovery, its contributions to national and global politics and culture, and its inevitable controversies and scandals.

Fossil Plants as Tests of Climate

The Sedgwick Prize for the best essay on a geological subject was instituted in memory of Adam Sedgwick, the geologist who introduced Darwin to geology in walking tours of north Wales, but later opposed his theories. One of its most eminent winners was A. C. Seward (1863-1941), then a young lecturer in botany at Cambridge. He combined the study of botany with geology in his research on what the age and location of fossilized flora can reveal about the climates of different geological periods. The author of the standard early twentieth-century textbook in the field, Fossil Plants for Students of Botany and Geology (1898-1919), he served as Professor of Botany at Cambridge, Master of Downing College and Vice-Chancellor of Cambridge University. This Sedgwick Prize essay sets out the state of knowledge in the field in 1892 and was the foundation of a lifetime's work in palaeobotany.

Cambridge University Press has long been a pioneer in the reissuing of out-of-print titles from its own backlist, producing digital reprints of books that are still sought after by scholars and students but could not be reprinted economically using traditional technology. The Cambridge Library Collection extends this activity to a wider range of books which are still of importance to researchers and professionals, either for the source material they contain, or as landmarks in the history of their academic discipline.

Drawing from the world-renowned collections in the Cambridge University Library, and guided by the advice of experts in each subject area, Cambridge University Press is using state-of-the-art scanning machines in its own Printing House to capture the content of each book selected for inclusion. The files are processed to give a consistently clear, crisp image, and the books finished to the high quality standard for which the Press is recognised around the world. The latest print-on-demand technology ensures that the books will remain available indefinitely, and that orders for single or multiple copies can quickly be supplied.

The Cambridge Library Collection will bring back to life books of enduring scholarly value across a wide range of disciplines in the humanities and social sciences and in science and technology.

Fossil Plants as Tests of Climate

Being the Sedgwick Essay Prize for the Year 1892

Albert Charles Seward

CAMBRIDGE
UNIVERSITY PRESS

CAMBRIDGE UNIVERSITY PRESS

Cambridge New York Melbourne Madrid Cape Town Singapore São Paolo Delhi

Published in the United States of America by Cambridge University Press, New York

www.cambridge.org
Information on this title: www.cambridge.org/9781108004275

© in this compilation Cambridge University Press 2009

This edition first published 1892
This digitally printed version 2009

ISBN 978-1-108-00427-5

FOSSIL PLANTS

AS TESTS OF CLIMATE.

London: C. J. CLAY AND SONS,
CAMBRIDGE UNIVERSITY PRESS WAREHOUSE,
Ave Maria Lane.

Cambridge: DEIGHTON, BELL AND CO.
Leipzig: F. A. BROCKHAUS.
New York: MACMILLAN AND CO.

FOSSIL PLANTS

AS TESTS OF CLIMATE

BEING THE SEDGWICK PRIZE ESSAY FOR THE YEAR 1892

BY

A. C. SEWARD, M.A., F.G.S.,

ST JOHN'S COLLEGE, CAMBRIDGE,
LECTURER IN BOTANY IN THE UNIVERSITY OF CAMBRIDGE.

" Effects similar in kind to those produced now must in all former times have been produced by some corresponding power of Nature."

(SEDGWICK, 1830.)

LONDON :

C. J. CLAY AND SONS,

CAMBRIDGE UNIVERSITY PRESS WAREHOUSE,

AVE MARIA LANE.

1892

Cambridge:
PRINTED BY C. J. CLAY, M.A. AND SONS,
AT THE UNIVERSITY PRESS.

PREFACE.

A FULL treatment of the subject "Fossil plants as tests of climatic changes," would require a more thorough acquaintance with fossil floras of various geological ages, as well as a wider knowledge of the distribution and conditions of life of recent plants, than I can lay claim to. Moreover the time at my disposal during the period allowed for the completion of the Essay was inadequate for a revision of the fossil floras, and a careful analysis of their geographical distribution. I have therefore confined myself to the endeavour to bring together such botanical and geological facts as seemed to me likely to prove helpful in climatal retrospects, calling especial attention to the several points of view from which previous writers have considered the subject. Various matters, bearing directly or indirectly upon the question, are briefly dealt with in the hope that such a summary may suggest methods for further enquiry, and facilitate the filling in of details in certain branches of the subject, the outlines of which I have roughly sketched.

I desire to express my hearty thanks to Professor T. McKenny Hughes and Professor W. C. Williamson for their advice and generous help, and also gratefully acknowledge suggestions received from Mr Carruthers and Mr F. Darwin.

A. C. S.

CAMBRIDGE, 1892.

TABLE OF CONTENTS.

TABLE OF CONTENTS

INTRODUCTION.

IF we take up any work treating of fossil plants we shall generally find the difficulties of the subject emphasized and the disadvantages recounted under which the palæobotanist is placed as compared with the palæozoologist. The latter finds rich material preserved in media, and under conditions, in which fossil plants never occur. Animals too, from the nature of their organisation, are far better adapted for preservation in a fossil state than the more delicate and perishable structures of plants. In questions of geological chronology and correlation animals have generally been preferred to plants, both on account of their greater abundance in the stratified rocks, and because they have been considered safer guides. Bunbury[1], writing on the vexed question of Indian stratigraphy, remarked—"The palæobotanical evidence is far from unequivocal, and, such as it is, might be outweighed by the discovery of a single, well marked and thoroughly characteristic, fossil shell or coral." This shews what little importance is attached, by some at any rate, to the correlative value of fossil plants.

Palæobotanists have for the most part expressed their confidence in plant evidence. We are not concerned, at present, with the defence of Palæobotany as a valuable aid to Palæozoology in stratigraphical geology, nor with supporting the opinion of some of the more sanguine that plants may indeed be more reliable tests of age than fossil animals. We have to consider plants as the thermometers of the past: in this light they have been regarded by most writers as more trustworthy guides than animals. Plants are unable to migrate with the same ease as animals when the temperature of their station becomes unfavourable. When the

[1] Feistmantel (3).

conditions become critical, they must either give way before the
stress of circumstances or gradually-adapt themselves to the altered
environment. As tests of climate, therefore, the first place has
generally been assigned to plants. The same difficulty is encountered in making use of fossil plants as tests both of geological
age and climate; namely, the scarcity and fragmental character of
their remains in sedimentary rocks, due to the comparatively
narrow limits of the conditions under which samples of geological
floras are likely to be preserved. Leaves, twigs and fruits of plants
blown or floated into the quiet waters of a lake or inland sea, make
up a large proportion of the material with which we have to deal.

Beds of peat, lignite, jet and coal supply us with abundant
fragments of land and swamp vegetation belonging to different
geological periods.

One of the most fruitful sources of error in the determination
of fossil plants has been the fragmentary nature of the specimens:
this is especially noticeable in the descriptions of the earlier palæobotanists—nor indeed is the same remark altogether inapplicable
to the writings of modern palæobotanists—in which determinations
have been based on imperfectly preserved casts or impressions and
have led to many and serious errors. The mistakes, however, must,
in very many cases, be put down, not so much to rashness or care-
lessness on the part of the workers, but to the imperfections of the
materials at their disposal. With the use of the microscope by
such pioneers as Sprengel, Witham and others, a new impetus was
given to Palæobotany; an impetus which established novel and
more reliable methods of research. With the help of the microscope we have been able to examine, in minutest detail, the structure of what in all probability is a Silurian Alga[1], and to detect, as
van Tieghem[2] asserts, Bacillus amylobacter in the act of eating
away the tissues of Carboniferous trees.

Binney, Williamson and Carruthers have made good use of the
rich material supplied by the Upper Carboniferous strata of Lan-
cashire and Yorkshire, while on the Continent Göppert, Stenzel,
Renault, Felix, and other investigators, have brought to light, by
the microscopic methods of examination, facts as important as they
are interesting.

The result of this microscopical method has been to place the

¹ Carruthers (1).   Penhallow (1).   ² van Tieghem (1).

science on a much firmer footing and—a fact of great importance
—to attract the attention of botanists to the field of Palæo-
phytology as one likely to afford a rich harvest in questions of plant
phylogeny.

The labours of Heer, Ettingshausen, Newberry, Starkie Gardner
and many others have shewn us of what importance a mere external study of plant fragments may be, both from a geological and
botanical standpoint. In the absence of those characters upon
which botanists rely in the determination of genera and species
the palæobotanist has to a large extent to content himself with
such features as would be deemed insufficient in living plants.
Ettingshausen[1] has endeavoured to shew how far nervation may be
depended upon in plant determination in the case of Phanerogams
and Ferns. As regards Ferns, their mineralized-tissues and, in some
few cases, fossil sporangia, have enabled the palæobotanist to speak
with certainty as to families and genera; but, as Stur[2] has shewn,
nervation alone when applied to Ferns cannot be depended upon
for purposes of classification. In the case of Phanerogams the
phytologist is still more dependent upon nervation, and those who
have given most attention to the subject contend that this test is
one which, used cautiously, gives satisfactory results.

On the other hand botanists have often loudly protested, frequently with just cause, against the extent to which nervation has
been used as a means of determining genera and species from the
merest fragment of leaves.

The loose methods employed by the earlier writers, combined
with the scarcity of good specimens, gave rise to the creation of a
large number of genera and species which more extended knowledge has shewn to have been in many cases founded upon
fragments of one and the same plant. One has only to look at the
catalogue of Palæozoic plants in the British Museum[3] to see into
what a state of confusion the synonymy of fossil plants has fallen.

Of the many sources of error open to the palæobotanist sufficient has been said by various writers: the remarks on this head by
Hooker[4] are especially valuable in the case of Vascular Cryptogams
and other plants. As regards the practice of identifying leaves, or
fragments of leaves, from their nervation, Bentham[5] has shewn the

[1] Ettingshausen (1) (2).   [2] Stur (1).   [3] Kidston (1).
[4] Hooker (1).   [5] Bentham (1).

dangers which attend such a method. It has recently been pointed out by Fontaine[1], in his monograph on the Potomac flora, that certain specific names given to fossil Ferns should not be considered to imply that all Ferns bearing the same specific name are really the same species. In certain cases the specific name means rather a type of venation, than a distinct species in the modern sense of the term. He quotes Sphenopteris Mantelli and the widely distributed Pecopteris Whitbyensis as examples of specific names which refer rather to a particular kind of foliage and nervation than to well-defined species.

With the object of finding out the different points of view from which fossil plants have been regarded as tests of climate, we shall notice such references as have been made by various writers on Palæobotany to the general question of plants and their value as indices of past climatal changes. This historical sketch is by no means complete ; our object being to discover what views have been held as to the application of Palæobotany to questions of geological climates.

After this retrospect of the different expressions of opinion, a number of facts are recorded bearing upon the distribution of plants with special reference to Arctic floras, also facts connected with the effect of external conditions upon plants as expressed in their form and minute anatomy. The Carboniferous flora is treated at some length, as it is a subject around which considerable interest has centered ever since fossil plants were thought worthy of serious study ; and, judging by recent expressions of opinion, there seems no little evidence throwing doubt upon views generally held on the distribution and character of Coal-measure vegetation.

The fossil floras of Arctic lands are briefly noticed as affording a striking instance of the application of plant remains to climatological questions, and as an example of a method of enquiry which may be expected to yield fairly trustworthy results.

---

[1] Fontaine (2), pp. 337, 338.

CHAPTER I.

HISTORICAL SKETCH.

IN the fifth Annual Report of the United States Geological Survey (1883—84) we have an elaborate and exhaustive *Sketch of Palæobotany* by Mr Lester F. Ward[1]. The early history and development of the science receives special attention. He reviews the several stages in the evolution of Palæobotany out of the mass of confused and extravagant theories which characterised what he has called the "Pre-Scientific period." Without attempting to travel back along the lines which have been followed in the slow and tedious growth of scientific ideas as to the true value and significance of fossil plants, we must confine ourselves to noting the growth of such theories or opinions as have been formulated with regard to the connection between fossil plants and climatic changes in the past history of the earth.

Here and there on the line of advance we find the works of pioneers of the science standing out as landmarks; at such we shall glance in passing and discover what views have been held from time to time as to the evidence which fossil plants afford in questions of geological climates. Discussing the writings of the eighteenth century, Mr Ward shews how geological considerations were utterly ignored in descriptions of fossil plants. "The investigators of the last century," he remarks, "were really not discussing the geologic age of fossil remains. The assumption was universal that these were plants that grew somewhere in the world only a few thousand years ago at most, plants such as either grew then in the countries where their remains were found, or in other countries from which they had been brought by one agency or another, generally that of the flood, or else, as some finally

[1] Ward (1).

conceived, had been destroyed by these agencies, so as to have no exact living representatives[1]." The theories held by writers of this period Mr Ward classes under three heads—(i) the indigenous theory, (ii) the exotic theory, and (iii) the extermination theory. The exotic theory is the only one which we need consider as bearing more directly than the other two upon the present subject.

Here we find the first suggestion of a relationship of European fossil plants to genera still living in the tropics. Although very far from the truth the upholders of this theory had made a decided advance. It was realised by some of the writers of this period that we are not compelled to recognise in fossils petrified remains of vegetation exactly corresponding to that growing in the districts where the fossils are found.

In 1706 Leibnitz[2] puts forward the suggestion that certain impressions of plants found in Germany agree most closely with species living in India.

In 1718 Antoine de Jussieu[3] in describing some Coal plants from Saint Chaumont considers them closely allied to living species from the East Indies and other tropical regions.

Several other writers are mentioned by Mr Ward as supporters of the exotic theory. Walch is especially worthy of note as having recognised in fossil plants from England, France and Germany a much closer agreement than is the case with the living floras.

To pass over the later decades of the eighteenth century we find a prominent landmark in the writings of Schlotheim[4]. He refers to Jussieu's opinion as to the occurrence of southern plants in our own latitudes and goes on to state, as the result of his examination of plants from the shales above coal seams, that they are products of southern latitudes; the Ferns shewing a close resemblance to East Indian and American forms.

The European Coal-measure fossils are spoken of as extinct species, and, from their supposed relationship with tropical forms, and the presence of Tree-ferns, the vegetation of that period is considered to have been of a luxuriant and southern character.

No important observations are made by Parkinson[5] in his classic work *On the Organic Remains of a former World*, so far as

[1] Ward (1), p. 395.
[2] *Histoire de l'Academie royale des Sciences;* année 1706, p. 10. Paris, 1706.
[3] de Jussieu (1).   [4] Schlotheim (1).   [5] Parkinson (1).

the question of former climates is concerned. He quotes the opinion of Dr J. E. Smith (President of the Linnean Society) that such genera as Cyclopteris, Pecopteris &c. from the Coal-measures are all foreign forms and from warm climates. Steinhauer[1] compares a sandstone cast of Calamites, which he names Phytolithus sulcatus, to a tropical Bamboo, but does not venture to discuss the general question of climate.

In 1825 Artis[2] brought out his valuable descriptions of fossil plants, and, in speaking of the difficulties attending the identification of fossil forms, lays stress on the necessity of studying "the whole anatomy of the plant"; he refers to the work of Martius who instituted a comparison of certain fossil stems with recent Brazilian forms. We come next to the most important stage of palæobotanical history. Early in the present century we find a number of most important and scientific writings, which we may regard as the first contributions to true palæobotanical science. Adolphe Brongniart, whom we may call the Father of Palæobotany, contributed in 1822 his first memoir on fossil plants. In 1828 appeared his well known *Prodrome d'une histoire des Végétaux Fossiles*[3]. We may consider briefly such points as bear more or less directly upon the question of climate.

In the introductory remarks Brongniart notices the fact, observed by several earlier writers, that there is but little analogy between fossil plants and those still living in European countries.

Antoine de Jussieu is mentioned as one of the first to suggest a similarity of fossil plants with recent tropical genera. Mr Ward, in quoting Antoine de Jussieu's views, adds in a foot-note, "It is remarkable that both Brongniart (Hist. des vég. foss., Tome I. p. 3) and Schimper (Traité de pal. vég., Tome I. p. 4) should have committed the error of crediting this paper to Bernard instead of Antoine de Jussieu." In the reference to Jussieu in the "Prodrome" to which I have referred[4], Brongniart speaks of Antoine de Jussieu and gives the reference to his paper in the "Mémoires de l'Academie des Sciences" for the year 1718. In the "Hist. des vég. foss.[5]" the reference is given to Bernard—as Mr Ward points out—and with the date 1708.

To return to the "Prodrome." After describing the present

---

[1] Steinhauer (1).   [2] Artis (1).   [3] Brongniart, Adolphe (2).
[4] Brongniart, Adolphe (2), p. 1.   [5] Brongniart, Adolphe (3).

distribution of Algæ the author proceeds to draw conclusions from a comparison between their past and present distribution. He points out the occurrence in strata anterior to the Chalk of species now confined to tropical zones, such as Sargassum, species of which are found in the rocks as far north as Sweden, whereas now no species occurs living beyond latitude 43° N. The same fact is noticed in the case of Caulerpa. The value of such a method will be discussed when we summarise the opinions and lines of argument of the several authors whose conclusions we are passing in review.

In describing the geological floras, Brongniart divides them into four main groups or periods. These periods are separated from one another by a sharp line and no gradual passage exists from one flora to another. The *first period* extends from the beginning of the "Transition" rocks to the end of the Coal period, the Zechstein forming the upper limit. The *second period* corresponds to the time of deposition of the Bunter rocks.

The *third period* begins with the deposition of the Muschelkalk and extends to the Chalk.

The *fourth period* embraces all strata above the Chalk.

In chapter II. the distribution of fossil plants in the different strata is fully discussed.

Under the head of "Terrain Jurassique" are included rocks ranging from the upper Lias to the Greensand. The proportions in which the several families of plants are represented are much the same as in the preceding "Terrain," whose lowest rocks are of Keuper age. Thus, from the close of Bunter times to the commencement of the deposition of the Chalk, the floras were characterised by the existence of only two of the great classes of the vegetable kingdom ; and the classes which were then predominant occupy at the present time a very subordinate position. These two classes Brongniart calls "Cryptogames Vasculaires" and "Phanérogames Gymnospermes." The predominance of Cycads in Europe is remarked as affording a striking contrast to their present distribution. The Ferns differ very considerably from those of the Coal-measures and Bunter, and cannot be considered as remnants of the Coal vegetation. The difference between the Liassic and the Coal floras is made still more striking, Brongniart goes on to say, by the recent observations of Élie de Beaumont according to whom strata occur in the Anthracite formation of Tarentaise, whose fossil shells denote a Liassic age but whose plants agree with Coal-measure forms. From this Brongniart concludes that we must suppose the existence in Liassic times of two sets of plants, the majority of those found in Europe being remnants of a vegetation which grew in temperate regions, others, such as those recorded by Élie de Beaumont, having been brought from other latitudes where plants characteristic of the Coal age still lingered.

A long list of plants is given from the "Terrain Houiller." The importance is recognised of not only comparing the past and present distribution of the various families, but of comparing the proportions in which the different classes of plants were represented in Carboniferous times, and the positions which the same classes occupy in the present vegetation. The following table shews the results arrived at[1].

Period	I.	II.	III.	IV.	Present
Agames	4	5	18	13	7,000
Cryptogames celluleuses	,,	,,	,,	2	1,500
Cryptogames vasculaires	222	8	31	6	1,700
Phanérogames gymnospermes	,,	5	35	20	150
———— monocotylédones	16	5	3	25(?)	8,000
———— dicotylédones	,,	,,	,,	100(?)	32,000
Végétaux de classe indéterminée	22	,,	,,	,,	,,
[2] Total of each Flora	264	23	87	166	50,350

The uniformity of the Coal vegetation in different latitudes is noted, and its two essential characteristics Brongniart considers to be, (i) the large proportion of Vascular Cryptogams, (ii) the great development of plants of this class. The climate which obtained during the Coal period, he considers was probably as warm as, if not warmer than, that of the tropics. Referring to observations of R. Brown and d'Urville, Bronginart discusses in detail the relative proportions of Ferns and Lycopods in the present floras: finding that these classes occupy the most important positions, as regards

[1] Brongniart, Adolphe (1).
[2] cf. Ward (1) pp. 440—441. (Table compiled from material collected up to 1883.)

relative proportion, in the floras of such regions as the Antilles,
St Helena and Ascension Island, the conclusion is drawn that at
the time of the formation of our coal seams insular rather than
continental conditions obtained : the temperature was tropical and
the atmosphere highly charged with moisture. The nature and
mode of formation of coal are questions intimately associated with
climatic considerations : these are not passed over by the author of
the " Prodrome."

Théodore de Saussure's[1] experiments are quoted as shewing that
the present percentage of carbonic acid gas in the atmosphere is
not the most favourable to plant life. Allowing a higher percentage in Upper Carboniferous times, we should have conditions more
favourable both for greater activity in the growth of plants, and for
the accumulation of vegetable *débris* destined to form seams of
coal by a process analogous to the growth of peat. This would be
rendered possible by the higher percentage of carbonic acid gas, as
that would offer a check to the decomposition of the dead plants,
which the high temperature would otherwise accelerate, and so
prevent such accumulations as would in time give rise to coal
seams.

Passing from the Carboniferous to more recent periods we have
evidence of a diminution of temperature. In the Cretaceous flora
we find characteristics of continental rather than of insular vegetation ; the climate a little warmer than that of central Europe
at the present day. Local influences had begun to make themselves apparent in the Cretaceous flora ; floras of different latitudes had begun to be marked out by characters dependent on
various external conditions. The two periods intermediate between
the Coal period and the Cretaceous probably enjoyed warm climates.
The Bunter plants are few in number, but the presence of a Tree-fern
suggests a temperature in Europe higher than that of our latitudes.

In the third period the vegetation possessed characters allied
to those which are presented by the floras of equatorial coasts and
large islands. The consideration of the present distribution of
Cycads leads Brongniart to this conclusion.

Before entering upon the discussion of the character and distribution of the vegetation of the four periods as compared with

---

[1] Brongniart, Adolphe (2), p. 186.

recent floras, Brongniart asks the question whether it would be possible to determine the conditions under which each flora lived from an examination of the characteristic fossils of each period. A botanist, he says, is able to discover the conditions of climate under which a given set of plants grew by examining the species. In the case of fossils we may also arrive at results which, if not certain, at least may be regarded as very probable.

In a review of Brongniart's work which appeared in 1829, Hoffmann[1] draws attention to the supposed existence in Liassic times of two regions of vegetation. In one region, according to Brongniart, was included Europe and probably most of our temperate zone; there the vegetation differed essentially from that which characterised the Carboniferous period : in the other region, corresponding to our warmer zone, there remained plants unchanged since Coal-measure times. The fact of Coal plants being found in temperate latitudes is explained by Brongniart by transport from a warmer and more southern district. This explanation is regarded by Hoffmann as unsatisfactory: he refers to other cases in which a supposed transport from distant latitudes has been called in to explain the occurrence of plant and animal remains in unexpected places. Hoffmann concludes that we have no evidence for the existence of zones of temperature in Liassic times. We should notice that Brongniart did not allow himself to fall back upon the theory of transport from other latitudes except in this particular case of the Swiss Lias strata discovered by Élie de Beaumont.

In 1821 a paper was contributed to the *Annales des Mines*, by Alex. Brongniart[2], in which we find it stated that the upright positions of some stems which had been found in the French Coal-measures did not prove that the trees had been preserved in their position of growth, but rather pointed to drifting; they had, however, not been carried any great distance, and, in Brongniart's opinion, their mode of occurrence could not therefore be considered in favour of the hypothesis that the Coal plants had been transported into our latitudes from the tropics. In the later works of Adolphe Brongniart we find the same opinion expressed on matters bearing on geological climates[3]

As in the case of Brongniart so with Sternberg[4], whose work

[1] Hoffmann (1).   [2] Brongniart, Alex. (1).
[3] Brongniart, Adolphe (3).   [4] Sternberg (1).

we may conveniently consider next, his writings extend over a number of years, and form another of the most conspicuous landmarks in the advance of Palæobotany.

Sternberg deals very fully with the results arrived at by previous workers, and in the main agrees with Brongniart in his deductions relating to evidence afforded as to climatal conditions in the four great periods of vegetation.

As usual the flora of the Coal-measures takes up the main portion of the work. This is compared to the vegetation in the region of an inland sea; on the banks were grasses and reeds, on islands and hills trees, shrubs and ferns. Here we notice a recognition of the fact that the plants found in one set of strata may have grown under different conditions, although in the same latitude. The occurrence of characteristic Coal-measure plants in China, Japan, Siberia, Newfoundland, Greenland, North America, Bear Island, &c., is cited as favouring the view of a uniform Carboniferous vegetation. So far as comparison of Coal plants and recent forms is possible, the conclusion arrived at is that the former must be considered tropical: it is pointed out that they cannot have been carried far by water before deposition, and may therefore be taken as fairly representative of the vegetation which grew in the districts where the fossils are found. Sternberg, in admitting a uniform climate during the Coal period, explains that a uniform temperature does not mean the same reading of a thermometer in all parts of the world, but rather that such places where the fossil plants have been found had " Plant isotherms," where plants grew which were either identical or closely related. The temperature of these isotherms was equal to, or possibly greater than, that of the tropics. The older floras point to higher temperatures and a greater extent of water on the surface of the globe: such, according to Sternberg, is the opinion of most geologists.

In a paper *On the climate of the antediluvian world and its independence of solar influence: and on the formation of granite*, by Sir Alex. Crichton[1], we have a number of noteworthy observations on the use of animal and plant fossils as tests of antediluvian climates. The evidence obtained by an examination of the character and distribution of fossils is in favour of a high and uniform

---

[1] Crichton (1).

temperature from high northern to corresponding southern latitudes. Valuable cautions are given against drawing conclusions from facts which at first sight might appear to afford unimpeachable evidence. In the case of the Mammoth bones found in northern climates the warning is given that elephants are animals of a migratory disposition, and there is further a possibility of transport after death. Another caution is necessary in dealing with the distribution of shells. Fossil shells in northern countries may be very like those now living in India and the South Pacific, but it must be remembered that Indian shells also occur in temperate latitudes. Such pieces of evidence as these may only be used therefore as concomitant proofs, and of themselves are insufficient to warrant any safe conclusions as to past climates. Thus far animal fossils have been dealt with. We are not directly concerned with fossil animals in the present discussion, but it is interesting to notice that the necessity of caution was recognised so early as 1825 in attempting to draw inferences from a comparison of fossil and recent animals as to conditions of climate. The same remarks might be made with equal point in the case of fossil plants.

Discussing the Coal flora Crichton remarks, " Every coal country in every part of the world which has hitherto been examined, abounds in the fossil remains of similar vegetables." The same writer lays stress on the aid likely to be afforded in the solution of climatal questions by an examination of fossil plants. " The laws of vegetable life," he adds, " as relating to temperature are positive, and therefore, when connected with the individuals of the antediluvian vegetables, they throw the greatest and surest light upon the subject of its climates."

Instances are given in support of the assertion that the similarity of any two floras depends more upon a similarity of temperature than of soil. Another cause, in addition to uniform temperature, is suggested in explanation of the wide geographical range of the older fossil plants. During the formation of coal there was less dry land than at present, and the waters would serve as a vehicle for distributing germs or seeds of antediluvian plants. Here is an important factor which must not be neglected when we come to describe in more detail the distribution of the Carboniferous flora. It is stated, as a natural consequence of an admission that plant-life was governed by the same laws in the earlier ages as now, that we must allow a greater uniformity of temperature to have obtained in the earlier ages of the world over the whole globe. The relative proportion of the different plant families in the older floras as contrasted with the positions which they occupy to-day is considered, and regarded as pointing to very high temperatures on the earth's surface in earlier times.

"At present," it is stated, "the ratio of Dicotyledons to Acotyledons and Monocotyledons increases in proportion to the distance from the tropics."

No Dicotyledons are recorded until Oolitic times and their absence in the older strata is taken as strong evidence that in those earlier days every part of the earth's surface was hotter than our hottest regions. Contrasting the relative value of plants and animals as criteria of past climates, the preference is given to plants, as they are not endowed with the same powers of locomotion as animals, and are thus unable to escape so easily from unfavourable conditions of climate.

In the paper, from which I have quoted at some length, many valuable suggestions are given which may with advantage be borne in mind in any attempts which are made to apply our more extended and accurate knowledge of Palæobotany to the question of past climates.

At the conclusion of this important contribution we find a recurrence to prescientific ideas when dates come to be considered. The fossils of the London Clay indicate a West Indian climate : this deposit was probably formed, we are told, about 6829 years ago : during that period, therefore, the climate of Great Britain has been reduced from the heat of the West Indies to its present standard !

Dr Boué[1] in 1826 contributed a paper to the *Edinburgh Philosophical Journal, On the changes which appear to have taken place during the different periods of the earth's formation in the climate of our globe, and in the nature, and the physical and geographical distribution of its animals and plants.*

"The distribution of Animals and Vegetables," says Boué, "is most materially influenced by the division of the surface of the Earth into zones, and into countries or climates." The successive establishment of these different zones and climates explains the difference in the animal and vegetable kingdoms in present and past ages. The further we penetrate into the crust of the earth the more simplicity is found in the vegetable and animal productions, and at the same time a greater uniformity. These facts are explained on the supposition of greater equality of temperature, and an absence of climatic zones in the earlier geological periods. In a paper in the same Journal two years later (1828), an important addition is made by the Rev. J. Fleming[1] to the literature bearing on climates and fossil plants and animals. A wholesome caution is given to those who attach an undue weight to arguments as to former climates based on analogy.

The following points are considered:

1. If two animals resemble each other in structure, will their habit be similar?

2. If two animals resemble each other in external appearance, will their habit be similar?

3. If two animals resemble each other in form and structure, will their physical and geographical distribution be similar?

Various examples are quoted answering these questions in the negative. "Every species is controlled by its own peculiar laws." Applying the same reasoning to plants, Fleming goes on to suggest that this country may have had its Palms, its Cacti, and arborescent Ferns, under a temperature similar to the present, due regard being had to species not genera.

Buchanan[2] in 1829 submits additional evidence for greater uniformity in the temperature of the earlier periods; the wide distribution of Cycads and Equisetum columnare being taken as evidence of this in Lower Jurassic times. Differences in climate probably first became apparent, he suggests, after the Deluge.

During the years 1831 to 1837 there were issued the several parts of the *Fossil Flora of Great Britain*[3], a work which will always rank first, in spite of the various corrections and modifications rendered necessary by more complete knowledge, in the list of our earliest English contributions to Palæobotany.

In the Preface, the authors begin by shewing the importance of the study of fossil floras in throwing light upon past changes in the earth's climate, and upon the question whether there has been

---

[1] Fleming (1).   [2] Buchanan (1).
[3] Lindley and Hutton (2).

a progressive development in the organisation of plants. A short
but graphic description is given of the conditions which obtained
during the coal formation: in the "most dreary and desolate
northern regions of the present day" the vegetation was tropical
and the air too highly charged with carbonic acid gas to admit of
the existence of the present animal world. Lindley and Hutton,
following the views of Brongniart, describe the vegetation at the
time of the Chalk as extra-European and chiefly tropical. "Immediately succeeding the Chalk a great change occurred and a
decided approach to the flora of modern days took place in some
striking particulars."

"In Fossil Botany, the evidence afforded in geological enquiries
is in the present state of the science, as nothing compared to what
is to be expected from future discoveries." After thus expressing
themselves the authors proceed to the discussion of the use and
probable misuse of negative evidence. Such calculations as were
made by Adolphe Brongniart, based on the proportions borne by
one class of plants to another in a given formation, are regarded as
untrustworthy. To emphasize this protest against Brongniart's
method Lindley adds a note in Volume III.[1] "Upon the value of
numerical proportion in the ancient flora of the world, with reference to a determination of climate." From various considerations
Lindley was "led to suspect that possibly the total absence of
certain kinds of plants, the as constant presence of others, and
several other points of like nature, might be accounted for by a
difference in the capability of one plant beyond another of resisting the action of water."

A table is given shewing the result of keeping immersed for
rather more than two years 177 specimens of various plants.
Briefly stated, the conclusion drawn from this experiment was as
follows: Dicotyledonous plants quickly decompose: Coniferæ,
Cycadeæ and Ferns effectually resist the decomposing action of
water; in the case of Ferns the fructification rots away.

Lindley concludes his account of the experiment in the
following words:—"Hence the numerical proportion of different
families of plants found in a fossil state throws no light whatever
upon the ancient climate of the earth, but depends entirely upon

---

[1] Lindley and Hutton, loc. cit. p. 4, vol. III.

the power which particular families may possess, by virtue of the organisation of the cuticle, of resisting the action of the water wherein they floated, previously to their being finally fixed in the rocks in which they are now found."

To return to the Preface of Volume I. In reviewing the evidence for a tropical climate in England during the Coal period the questionable nature of the evidence is remarked. The presence in coal mines of Tree-ferns and Palms is one of the pieces of evidence whose value is overrated in temperature calculations. Palms, it is pointed out, occur in the south of Europe and in Barbary: a very moderate elevation of temperature, by no means tropical, would enable them therefore to grow in more northern latitudes. This is worth noting here, although we know that Palms have not been discovered in Palæozoic rocks. With regard to the supposed Tree-ferns Lindley and Hutton dissent from the views held by Brongniart, Sternberg and others that the Sigillarieæ should be referred to that class of plants. The shifting of the earth's axis is considered a reasonable explanation of the existence in former times of a rich polar vegetation. The theory of progressive development advanced by Brongniart is assailed with many arguments. Finally, methods of examination of fossil plants are dealt with at some length, and the importance of examining microscopically the minute anatomy of stems is insisted upon in such cases as the state of preservation of the fossils will permit. Witham is referred to as the first to adopt this method of examination.

Without entering into details, we may note in Witham's work the introduction of a new and extremely important factor in the elucidation of ancient climatal changes. Glancing at the result of this new method of microscopical examination as set forth in *The internal structure of fossil vegetables*[1], we may notice the aid afforded to the climate question by the presence, or absence— partial or complete—of annual rings in transverse sections of coniferous trees. Carboniferous Conifers shew imperfectly marked rings, resembling in this particular recent tropical trees. Witham considers that this leads to the legitimate inference that the seasons were probably not abrupt. The cells in the tissues of these older

[1] Witham (1).

Conifers were found to be larger than those in their modern representatives. Similar sections of Conifers from Liassic and Oolitic strata shewed more distinctly marked concentric rings, having the same irregularity in their structure as in the annual rings of living trees; then, as now, there was probably an irregularity in the successive seasons.

In 1843 Thomas Gilpin contributed *An essay on organic remains as connected with an ancient tropical region of the Earth*[1]. In this paper he brings forward evidence from animal and plant remains of a tropical zone having occupied in the earlier periods a different position to that of the present tropics. He considers, as usual, the Coal-measure plants to be relics of tropical life. The immense size of plants found in Carboniferous rocks is quoted as proof of tropical conditions.

The argument from size recurs again and again in writings dealing with the Coal flora; its weight will be discussed later.

In the monumental work of Göppert[2], published in 1836, entitled *Systema Filicum fossilium*, we find general remarks on climate questions. In the preface he dissents from Lindley's suggested explanation of the absence of Dicotyledons in the Coalmeasures. Göppert had previously accepted the experiment of Lindley, of which an account has been given, as affording valuable information which threw light on the apparent absence of certain plants from the Carboniferous flora.

The first part of this work is taken up with a very full history of Palæobotany, with special reference to fossil Ferns.

As an example of the application of a method which will be referred to later, we may take Göppert's work as a helpful model. The lines which he adopts in working out the distribution of fossil Ferns are such as might be followed in attempting to read past climate by the help of the distribution of fossil plants. After giving a summary of the distribution of Ferns in the several formations, Göppert shews how the tropical nature of the Ferns is apparent in the different periods. In no single case is there such a collection of genera and species as we find to-day in the temperate or northern zones of the world.

In Corda's standard work[3] on fossil plants we find certain

[1] Gilpin (1).   [2] Göppert (1).   [3] Corda (1).

genera, Diploxylon and Sagenaria, described as possessing large vessels, similar in size to those of their supposed living relations in the tropics, reminding us of Witham's remarks on the cells of Carboniferous Conifers. Corda considers this unusual structural development, as illustrated by the size of the vessels, a natural consequence of the higher temperature, and thicker atmospheric conditions which, according to Élie de Beaumont, characterised the coal period. The Ferns are chosen by Corda as most likely to throw light upon climatal changes : both recent and fossil forms having been well examined, and the temperatures recorded of each district where Ferns abound at the present day. The Marattiaceæ, for example, belong to a well defined zone of temperature, and are therefore well adapted for comparison with fossil forms in connection with climate questions. Marattiaceæ suggest a tropical temperature. In former times they were much more abundant than now, and seeing that the average temperature of such districts where the family is now represented is 25·8°C. we may conclude that in Upper Carboniferous and Lower Permian times the average temperature was the same. Sir Joseph Hooker [1], in his article *On the vegetation of the Carboniferous period as compared with that of the present day*, has given a valuable summary of the difficulties and sources of error in the determination of fossil plants. To quote Hooker's words, " We can never hope to arrive at any great amount of precision in determining the species of vegetable remains, nor to ascertain the degree of value due to the presence or absence of certain forms, such as the animal kingdom so conspicuously affords. Still less can we expect that they will prove equally appreciable indices of the climate and other physical features of that portion of the surface of the globe upon which they once flourished."

The necessity of a wide knowledge of recent plant distribution is pointed out ; and we are reminded how slight local causes may materially modify the operation of the laws of distribution as applied to living plants. Localities apparently occupying similar positions with regard to heat, light, soil and moisture, may be tenanted by genera of very opposite botanical characters. Ferns are specially mentioned as having a very wide range and being exceedingly sportive. Passing to Carboniferous plants the uniformity of their

[1] Hooker (1).

distribution is noticed in extra-tropical countries of the northern hemisphere.

The relation between plants and the nature of the soil is a point not hitherto noticed in reference to fossil plants. The fact of Sigillaria being characteristic of underclays and Conifers of sandstones suggests some connection between the plants and the material in which they are imbedded. The Conifers in sandstone were no doubt transported. Both Witham and Corda drew attention to the large size of the tissue elements in certain Coal plants. Hooker notes the succulent texture and great size of both vascular and cellular tissues of many of the Coal-measure plants, and regards them as possible indications of the presence of much moisture in the Coal period atmosphere. The presence of Coal plants in the arctic regions is considered proof of other conditions than the present having obtained in polar regions: the tissues of these Palæozoic plants being of too lax a nature to be able to withstand prolonged darkness and frost.

The vegetation of the Carboniferous period was certainly luxuriant, but not therefore necessarily varied. At the present day we are reminded that "in the temperate latitudes particularly, a recent flora, marked by a preponderance of Ferns, is also universally deficient in species of other orders." The preponderance of Ferns is legitimately adduced in proof of the temperate, equable and humid nature of the Coal-measure climate. Hooker adds we may further conclude the flora was poor in species, an inference warranted by the known facts of plant distribution.

Godwin-Austen[1], in a paper on the *Extension of the Coal-measures,* brings out the fact, often neglected by later writers, that the Coal-measure vegetation may be divided into upland and lowland. The lowland vegetation which covered the low-level surfaces must have been composed of dense growths of such plants as were capable of maintaining themselves, like the peat vegetation of modern times, for indefinite periods on the same spots. In Cyclopteris and Odontopteris in the sandstones of the coal basins Godwin-Austen recognises drifted vegetation of higher regions of the district.

In discussing the effect of distribution of land and water on climate, Hennessy[2] notices what appears to be contradictory evi-

¹ Godwin-Austen (2).   ² Hennessy (1).

dence afforded by the insects and other fossils from Liassic and Triassic rocks. In such cases as these Hennessy thinks the contradiction will be found to be only apparent, if we take into consideration the fact that temperature depends not only upon the height of the land above sea-level, but also upon the distance from the coast.

Dana[1], like Godwin-Austen and others, also notices a possible difference in habitat of certain Coal-measure plants. Conifers and Lepidodendra probably covered the plains and hills, and were confined to dry land; Sigillarieæ and Calamitæ grew in the marshes. The orthodox belief in a uniform genial climate is adhered to. The greater heat of the Carboniferous age, Dana suggests, would give rise to a greater amount of moisture in the atmosphere. The proportion of carbonic acid gas was also greater, and exceeded the present percentage by the amount stored away, in the form of carbon, in the coal of the earth[2].

Croll[3] offers a protest against the supposed tropical character of the Coal-measure climate, and considers it was rather moist, equable and temperate than tropical[4]. Robert Brown is quoted as supporting the same opinion; he considers the rapid and great growth of Coal plants suggests swamps and shallow water of equable and genial temperature.

Humboldt had previously shewn that Ferns abound most in mountains, growing in shady and moist parts of equatorial regions, and not in the hotter parts.

Without professing to adhere closely to chronological order we may notice some of the main points mentioned by Sir Charles Lyell with reference to our present subject. Referring to Hook's suggestion in 1688 of a warm climate indicated by fossil Turtles and Ammonites in the Portland Oolite, Lyell[5] regards the method of drawing deductions as to past climate in the older periods as impracticable, because fossil plants are specifically distinct from living plants, or even belong to different genera and families. If we go back gradually from the present we are more likely to escape the danger of considering climate the only cause, or even the main

---

[1] Dana (1).   [2] On this question see also Sterry Hunt (1) pp. 46—48.
[3] Croll (2).
[4] Croll (1), Geological climates, especially the more recent, are discussed in this latest work of Croll.
[5] Lyell (1), vol. I. p. 174.

cause, of the predominance of certain families without taking into
account the absence of competing forms of higher grade. This is a
possible source of error one must constantly keep in view in dealing with the evidence of past climatal changes. Attention has
recently been called to the same point by the late Prof. Neumayr[1]
in his lectures on changes in climate. Additional examples are
given by Lyell, illustrating the need of caution in reasoning from
analogy in the question of anatomical structure and its dependence
on external conditions of life.

Fleming and Cuvier are quoted in connection with this subject;
the former Lyell considers somewhat too sceptical, and believes
it is quite possible to judge of climate from collections of fossil
plants, even though the species be all extinct. In cases where
living and extinct forms occur together, the unknown becomes
known by the fact of its association with known species and can
afterwards be used as a standard of comparison. The importance
of bearing in mind the absence of competing forms of higher grade
is referred to in discussing the Carboniferous climate. The preponderance of Ferns in the Coal forests is by no means such an
important argument as Brongniart supposed; he did not take into
account the absence of competing forms. Lyell goes into some
detail in examining the data on which conclusions have been
drawn with reference to the climates of the several geological
formations. In the chapter on Botanical Geography[2] he quotes a
number of facts which bear more or less directly upon the subject of ancient climates. In speaking of the dispersion of plants,
Cryptogams are described as particularly well adapted for distribution; Lycopodium cernuum, for example, having a wide range in
equinoctial countries. This is a point not to be lost sight of in
considering the force of the argument so often used regarding the
uniform character of the Coal flora as proof of a uniformity in
temperature.

In another place Lyell[3] points out that we must look to Ferns
and Conifers as the best guides in helping us to conclusions as to
the climate of Carboniferous times. But Conifers, according to
Hooker, are of doubtful value as they occur alike in cold and
hot climates.

[1] Neumayr (3).   [2] Lyell (1), vol. II. p. 381.
[3] Lyell (2), p. 425.

In 1869 Schimper's *Traité de Paléontologie*[1] was published; another landmark in the progress of palæontological investigations. A section is devoted to the "application of Palæontology to Climatology." The climate of the Geological periods is considered, and a thermometric reading taken of the temperature of the Coal period is put down at 22° to 25° C. Ferns are considered very important tests of climate. The fact that there are fewer Ferns in North America than in central Europe is noted as a consequence of the much greater contrasts between the seasons in the former country.

Of primary importance in questions of climatology of the past are Heer's researches in Arctic and European fossil floras[2]. In a strictly chronological sequence Heer's work would have been considered earlier; we may, however, briefly examine here the general principles which he followed in making use of fossil plants as thermometers of ancient climates. Heer[3] divides fossil plants into three main groups:

1. Species which have only distant relatives in the present floras.
2. Species very closely allied to recent forms ("homologous species").
3. Species still represented in the modern floras. In the older formations the number of homologous species becomes smaller, and conclusions as to climate are therefore rendered more difficult.

The Carboniferous flora was uniform in character and affords no indication of a zonal difference in temperature. In Jurassic and Lower Cretaceous times there is still no suggestion of temperature zones. When we come to Upper Cretaceous strata the fossil plants afford proofs of difference in temperature depending upon geographical position, and this is still clearer in the case of Miocene fossils. In Switzerland Heer gives 23°—25° C. as the probable temperature in Coal-measure times; this probably remained unchanged, until the middle of the Cretaceous period: at the beginning of Lower Miocene times the thermometer registered 20·5° C. as the mean annual temperature, in the Upper Miocene

[1] Schimper, W. P. (1).   [2] Heer (1), (2), (3).
[3] See also Rothpletz (2). (Review of Heer's work in Obituary notice.)

18·5° C., in the Upper Pliocene 9° C., during the first Glacial period 8—9° C., during the second 4° C. and at the present time 9° C.

Correspondingly lower mean annual temperatures have been assigned to more northern latitudes; these will be considered in treating of Arctic fossil plants.

In the third volume of the *Flora tertiaria Helvetiæ*[1] a considerable space is devoted to questions of climate. Heer bases the conclusion that the Tertiary climate was warmer than the present on five chief arguments :

1. The greater number of species in Tertiary plants.
2. Predominance of woody plants.
3. Predominance of evergreen trees and shrubs.
4. General southern facies of the flora.
5. The flowering period of many Tertiary plants at the time of leaf development.

In considering such plants as are native in warm zones we see, as Heer remarks, that many extend into temperate latitudes, or can be cultivated in such latitudes, but he points out that, as Gaudin has shewn, we must enquire :

i. If the plants pass the winter and develope no seeds.
ii. Or if seeds, no further development.

Plants may be able to live in temperate latitudes, but be unable to form seeds, or else, if seeds are produced, to develope further.

In some general remarks on plants and climatology the same author expresses himself as follows[2]:—"If correct inferences can be drawn from existing animals and plants as to the climate of the country to which they belong, accurate conclusions may also be possible with respect to extinct or fossil species; and our deductions will attain more certainty in proportion to the abundance of the materials at our command, and to the near alliance of the animals and plants to those now living." Heer, as the result of his examination of the Swiss Tertiary plant beds, is led to the interesting conclusion that in certain cases it is possible to detect the regular recurrence of seasons by the constant association in the same

---

[1] Heer (2), vol. III. p. 327, "Rückschlüsse auf die klimatischen Verhältnisse des Tertiärlandes."

[2] Heer (1), p. 126.

strata of fruits or leaves of plants whose living representatives are known to agree closely in their periods of vegetation.

Some useful hints are given by Balfour in his *Palæontological Botany*[1]. He shews the importance of a wide knowledge of plant distribution in attempting to use fossil plants as indices of past climates. Fleming, Lyell and others have given instances of closely allied animals living under very different conditions of climate: Balfour quotes instructive cases of plant distribution. Palms are generally regarded as characteristic of warm countries, but, as Lindley and Hutton have previously pointed out, some species occur in temperate latitudes.

Chamærops humilis extends as far as latitude 43°—44° N. in Europe, C. palmetta of N. America grows in latitude 34°—36° N., C. Fortunei from the north of China is hardy in the south of England. Again, Palms and Bamboos are associated high in the Himalayas with Conifers.

Dr Haughton[2] expresses confidence in plants as tests of climate when he remarks—"We may regard the plants and animals found in the fossil state in the arctic regions as self-registering thermometers recording for us the mean temperature of those regions at successive epochs, marking so many fixed points on the earth's thermometrical scale."

The Marquis of Saporta, whose researches have thrown so much light upon the Geological floras of France, and who, following Heer's example, has used the floras as keys to past climatology, repeats the caution as to the value of extinct species in climatological questions. As we pass to extinct species we are reminded that the data on which the calculations as to climate are based become less precise.

As an example of methods followed in dealing with Tertiary floras we may note the chief points in Saporta's detailed description of the flora of Aix[3]. Saporta shews that on examining local and regional floras great differences are found in the relative proportions in which the two great classes of Phanerogams are represented. General humidity increases the proportion of Monocotyledons and diminishes that of Dicotyledons; a lowering of temperature has the same effect; a dry and warm

---

[1] Balfour (1).   [2] Haughton (1), p. 266.
[3] de Saporta (3).

country having a larger proportion of Dicotyledons than a warm
and moist or cold and moist country.

In the high and dry country of the interior at the Cape
(S. Africa), Monocotyledons make up 16·9 per cent., Dicotyledons
83·1 per cent. of the Phanerogams.

From the point of view of relative proportion in Monocotyledons and Dicotyledons, the flora of Aix is considered to indicate a warm and dry climate.

Saporta concludes from an examination of Tertiary floras that
herbaceous families of Phanerogams, which to-day are so common,
especially in Europe, occupied a more or less subordinate position.
One must admit for Eocene and Lower Miocene floras an enormous
preponderance of families of woody plants.

In the Aix flora the Leguminosæ occupy the first rank as in
most intertropical floras. Grasses make up 4 to 5 per cent. of
the Phanerogams, as in New Guinea. Similar comparisons are
drawn between the Aix and recent floras founded on the positions
occupied by different genera and families. The temperature in
Southern France during the Eocene period was probably more or
less tropical.

The abundance of Leguminosæ suggests a warm and dry
climate. Such plants as Musaceæ, Mimosa, Myrica, Diospyros and
others which are represented in the Aix flora do not now pass
the isotherm of 25° C. The preponderance of Palms, Araliaceæ,
Acacias, &c. points to a high temperature; with such types as
these occur others suggestive of cooler conditions, e.g. Microptelea,
which resembles M. Hookeriana found at a height of 4000—5000
feet in Sikkim.

A well-marked feature of the Aix vegetation is the feeble
development of appendicular organs, and the small size of many of
the plants compared with their living representatives. The leaves
of Araliaceæ, Magnolia, Sterculia, &c., cones of Pinus, fruits of
Paliurus tenuifolius, &c. are examples of this. Other characteristics are the coriaceous consistency of many leaves, frequent
spiny margins, great complication in the nervation of leaves, &c.
Leaves of a pliant consistency are rare.

On the whole the Aix flora may best be compared to that of
Africa, from Abyssinia to the Cape. Following the method
employed by Heer in his work on the Oeningen plants, Saporta

makes out three distinct sets of plants marked off from one another in their time of flowering. In the Aix beds such conclusions are less easily arrived at than in the Oeningen beds, because of the smaller size of the slabs on which the fossil leaves, flowers and fruits are preserved.

In the first group are included such genera as Microptelea, Populus, Laurus, Camphora, Pistacia, &c. which flower before the leaves are fully developed. Quercus, Ostrya, &c. make up the second group, in which flowers and leaves are formed at the same time. In the third group are Nerium, Magnolia (persistent leaved types), Diospyros, &c., plants which produce their new leaves before flowering. Indications are found of two distinct seasons, a dry and a rainy season.

Two maps are given; in one are shewn the present geographical limits of the principal genera represented in the Aix flora, in the other an attempt is made to shew the distribution of land and sea, and also the distribution of some of the more important types of vegetation during the latter part of the Eocene period. We may note that Saporta[1] in a later paper on the Provence floras sees no reason to change his opinion as to the climatic conditions which obtained in the South-east of France at the time of the Aix Eocene flora.

In the more recent plant-bearing strata there are not a few instances where plants, characteristic of different latitudes, are preserved side by side in the same rock. Such a commingling is considered by Mr Starkie Gardner[2] to be due to repeated cyclical changes in temperature, and the resulting migration of plants to warmer and colder latitudes. He remarks: "It must be borne in mind that it is not so much the mean temperature of a whole year which affects the possibility of plants growing in any locality, as the fact of what are the extremes of summer and winter temperatures." The bearing of such a commingling of species on climatal questions is of considerable importance. Another explanation of such a phenomenon has been suggested by Mr Searles Wood, Junr.[3] in the case of the Eocene plant beds of Hants. He suggested that the river, whose delta deposits contain the plant remains, flowed from the West through a district where a tropical climate obtained on the low ground, fed, like the Ganges, by

---

[1] de Saporta (4).   [2] Gardner (4).   [3] Wood, Searles (1) (2).

tributaries which flowed from a mountain region clothed with all kinds of vegetation, thus accounting for a mixture of tropical and northern forms in the delta sediments. Gardner rejects this view as untenable considering the state of preservation and mode of occurrence of the plants. In the section treating of the Carboni- ferous flora we shall find that Grand'Eury has paid special attention to the subject of climate and how far it is recorded in fossil plants.

In Phillips' *Manual of Geology*[1] we find little confidence placed by the author in the use of fossils as tests of climate. It is impossible, he says, to recognise climatic conditions from the structure and collocation of species.

Forbes has shewn that in the case of recent plants every species is controlled by its own peculiar laws; if then, he adds, one recent species cannot indicate the conditions of climate under which another species lived, much less can rare and surviving recent species tell us anything of the climatic adaptability of extinct fossil forms. To take an example: the Nipa fruits of Sheppey are of the same genus which now lives in the Ganges and Irawaddy district, but it is not possible to infer from this that the London clay climate was like that which now obtains in India.

Prof. Renault[2], in the introduction to his *Cours de Botanique fossile*, refers to plants as better guides than animals in questions of ancient climates. They are more susceptible to changes of climate, and have not the same means of migration should the conditions become unfavourable. Crichton long ago called atten- tion to this fact. The evidence from annual rings is considered worthy of notice; the zones of growth become more and more marked as we ascend the geological scale. The uniformity in size of the elements of the wood of Carboniferous trees is con- sidered proof of a regular climate: there is no evidence, according to Renault, for Coal-measure zones of temperature, the same plants being found in arctic and tropical regions. In the Creta- ceous period we meet for the first time with facts suggestive of a difference in floras dependent upon geographical situations. The zone of tropical climate is found to contract gradually as we pass from the Cretaceous period through different divisions of the Tertiary period.

[1] Phillips (1), p. 453.   [2] Renault (2).

Sir Archibald Geikie[1], in treating of the use of fossils in geology, alludes to the danger of relying on one species in climatal questions. We can only reason from analogy when dealing with an assemblage of fossils. In any case inference from analogy becomes more and more unsafe in proportion to the antiquity of the fossils and their divergence from existing forms.

In the second volume of the *Erdgeschicte*[2] Neumayr discusses the methods employed in attempting to determine former climates. Fossil forms are compared with their nearest living analogues: the arguments on which such comparisons are based depend upon the assumption that (1) closely allied forms must live under the same conditions, and (2) that no acclimatisation has occurred.

Examples are given shewing the insufficiency of the evidence derived from such considerations as these. To quote one: Foxes live in the coldest as well as in the hottest regions. .

In certain formations tropical plants predominate, but with them are occasionally found others now living in cold districts; for example, in the Chalk of Bohemia the Cherry, Willow and Ivy are found associated with a number of tropical forms. Cases are given where the evidence as to climate afforded by different families or genera is conflicting, demonstrating the widespread operation of acclimatisation. To quote another case from the animal kingdom: as Neumayr remarks, the occurrence of Ammonites is generally taken as proof of a warm climate. This is thought to be a legitimate conclusion considering the fact that the Nautilus, which is more or less closely allied to Ammonites, lives in tropical seas.

Again, size is no proof of a tropical habitat[3]: many large Cephalopods are found living in temperate zones. We may with advantage bear in mind this remark about size in dealing with the older fossil plants. It is pointed out that continental faunas and floras are more used to a struggle for existence than the plants and animals of islands, and may be considered, therefore, stronger and better able to hold their own in the face of competing forms. Whilst advocating the necessity of giving due allowance to acclimatisation as an important factor, not to be omitted in our attempts to read temperatures of past ages, Neumayr proceeds

---

[1] Geikie, A. (1), p. 613.   [2] Neumayr (2).
[3] See also Judd (1), p. 71.

to offer a warning against carrying such ideas too far. Acclimatisation goes on very slowly and makes the uncertainty of the method which relies on comparison of species more conspicuous in the case of the older formations. In more recent formations, where Tertiary plants are put in evidence, reliable information may be expected from the comparison method, but even here it is not possible to go so far as to determine actual temperatures. In the older formations the worth of this method is extremely doubtful; a much more trustworthy method, and one more likely to be fruitful in valuable results, is to work out the distribution of single types in the different formations, and thus obtain data affording an answer to the question—"Is there any evidence of the existence of climatal zones?"

In following out such lines of investigation one must be careful to notice the various causes which may operate in bringing about a specific difference in different latitudes; the distribution of land and sea, for instance, has an important effect in determining the range of species.

In Neumayr's own work we have an excellent example of such a method of enquiry as he advocates. Although he collects evidence from certain families of fossil animals we may sketch the arguments employed and the results arrived at, as affording a model of a new method of research, which might be copied when dealing with plants, and which might possibly yield results no less interesting and instructive than those furnished by Neumayr. We may summarise the more important points of his paper *Ueber klimatische Zonen während der Jura- und Kreidezeit*[1]. He begins with remarks on the prevailing idea that climate zones are first noticeable in Tertiary times, the pre-Tertiary warm and uniform temperature being erroneously explained by the supposed effect of the internal heat of the globe. He groups under three heads the arguments for a higher and more uniform climate in past time.

I. Luxuriance of vegetation as seen in the great extent of coal seams.

II. Geologically old organisms are more like living tropical forms than those inhabiting extratropical countries.

[1] Neumayr (1).

III. Floras and faunas are found to be similar in widely separated geographical districts.

I. To take the first argument: both Lyell and Croll have shewn that the conditions for the formation of coal seams would be more likely to obtain in temperate than in tropical countries; in the latter, decay would be too rapid to allow of sufficient accumulation of vegetable *débris*, and moreover peat, to which many coal seams may be best compared, is a formation characteristic of temperate or even cold countries.

II. Reef-building Corals are taken as a typical instance of the application of the second argument. They now inhabit seas where the temperature is not lower than 20° C.; Corals being found in Carboniferous limestone of northern latitudes, are assumed to have required the same temperature as their living representatives. Cephalopods are considered, on the same principle, evidence of warm climates: the Nautilus, however, being simply a survivor of a once widely-spread family, cannot be legitimately used as an index of the climatic conditions under which extinct Cephalopoda were able to exist. At the present time the largest Cephalopoda are found on the coasts of Newfoundland, Ireland, Norway and other northern regions, and not in the tropics. Cases are cited where representatives of fossil forms prefer a cold climate: in the Bryozoa, the Cyclostomata are decidedly arctic forms, and in the geological formations these are by far the commoner kinds. In any case it is important to remember that very often closely related types occur under widely different conditions of climate[1]; also we must bear in mind the great powers of adaptation possessed by animals, as exemplified in the Siberian Elephas and Rhinoceros. The occurrence of Cycads and Ferns in high northern latitudes in Mesozoic strata points to a warm climate, or at all events to a climate marked by the absence of severe frost.

In such a case as this one cannot get over the difficulty by supposing that the plants could have adapted themselves to an arctic climate such as now obtains in northern latitudes.

III. The "finds" of Carboniferous plants in latitude 75° N., like those from southern latitudes, are given in illustration of this argument.

[1] Prof. Judd, in his Presidential Address delivered before the Geological Society in 1888, quotes some interesting examples further illustrating this truth. Judd (1).

Neumayr deals at some length with the theories, so often quoted, which assign a larger percentage of carbonic acid gas to the Coal-measure atmosphere, and brings forward strong arguments against them.

If the atmosphere of the Coal period had been so rich in carbonic acid gas, the waters of the sea would have absorbed more and would have been able, therefore, to dissolve more readily calcareous sediment; under such conditions we should not expect to find great thicknesses of Carboniferous limestone, monuments of widespread and long-continued accumulation of calcareous material on the ocean floor. After bringing forward proofs, of distinct zones depending upon difference in latitude, as furnished by the facts of distribution of Jurassic Ammonites, Neumayr cautiously refrains from assigning definite temperatures to Jurassic climate; the data being too incomplete to admit of such temperature readings. Jurassic Corals extend as far north as England: at the present day the reefs of the Bermudas are the most northerly; hence, taking reef-building corals as guides, we should extend the present isotherms about 20° further north. If we take Jurassic Bryozoa, and make similar deductions from analogy with recent species, different results are obtained; hence it is better to reserve any decision.

Reptiles, Tree-ferns and Cycads are also taken as evidence for a warmer climate in the Jurassic age. Tree-ferns are found, however, in South America in quite cool places, and it is impossible to tell whether the necessary conditions of existence of any particular orders of plants have always been the same.

This detailed examination of marine mollusca leads Neumayr to define several parallel zones in Jurassic and Cretaceous times, these zones being the expression of differences in climate in different latitudes. From the beginning of the Jurassic to the Aptien period the same climatal conditions appear to have obtained. One zone extended round the equator; on the two sides of this was a northern and a southern zone respectively, these two zones being marked by the same zoological characters, despite their separation by 60° of latitude. Further north was a Boreal zone; want of data prevented the establishment of a corresponding Antarctic zone.

In a recent paper on the *Evolution of Climate*, by Prof.

James Geikie[1] we may notice a few points connected with the
general question of geological climates. He considers the conclusions of Neumayr hardly justify us assuming the existence
of climate zones in the Jurassic and Cretaceous periods. Such
differences in the fauna as are taken by Neumayr to be indicative
of climatal variations, are attributed by Geikie to such varying
physical conditions as depth of water or character of the sea-
bottom. Neumayr, we may remember, did not overlook the possible effect of such variations, and indeed recognised in the case
of some Ammonites the important part which the nature of the
sea-bottom may play in determining the limits of distribution.

Geikie concludes that on the whole the Mesozoic climate was
less obviously uniform than that of the Palæozoic era, but no
marked zones of climate had as yet been evolved. The microscopic structure of the wood of Palæozoic trees is taken as strong
evidence in favour of the absence of seasonal changes: in Cretaceous
Conifers regular annual rings are detected, which are not much less
marked than in recent trees.

Examples are given to shew that in the Cainozoic period a
well-marked differentiation of climate was established, which had
left unmistakeable evidence in the distribution of Tertiary floras.

In the foregoing historical sketch my object has been to take
a rapid survey of the views of some of the writers on the subject
of Palæobotany as connected with ancient climates. For a complete history of the development of Palæobotany we must turn to
Göppert's *Systema Filicum Fossilium*[2], or to Lester Ward's *Sketch
of Palæobotany*[3].[4] All that I have attempted is to notice the
different standpoints of various writers in discussing the question
before us. By analysing the arguments upon which conclusions
have from time to time been based as to past climates, we shall be in
a better position to answer the questions: (1) How far can fossil
plants be used as thermometers of the past? (2) What are the
best methods to adopt in attempting to apply Palæobotany to the
question of geological climates? The eighth law of Pictet holds
true—perhaps with less force—if the comparison he speaks of is

---

[1] Geikie, J. (2). [2] Göppert (1). [3] Ward (1).
[4] See also Knowlton (2). In the introduction to this paper a review is given
of the progress of the study of internal structure in Palæobotany.

made between floras instead of faunas: according to this law—
"La comparison des faunes des diverses époques montre que la
température a varié à la surface de la terre[1]."

After taking a retrospect of the lines of argument followed by
different writers who have discussed the value of plants as indices
of temperature, we shall proceed to treat in more detail some of
the methods of enquiry which may enable us to turn to the best
account such data as recent advances in Palæobotany have placed
at our disposal.

As far back as Walch, Ant. de Jussieu, Leibnitz, Parsons and
others we find the plants from the Coal-measures compared with
tropical vegetation; this idea has kept a firm hold on the minds
of many palæontologists up to the present time. When we
come to discuss the nature and distribution of Coal-measure
plants we shall endeavour to shew that the arguments in favour
of the tropical or subtropical nature of Palæozoic floras are by no
means so sound as they have generally been considered. To follow
the example of Brongniart, and attempt to base any conclusions on
such unsafe guides as Algæ would be excessively rash. Without
commenting at length upon fossil "Algæ" we may bear in mind
the remark of Hall[2]: "It has been the habit to refer to vegetable
origin all those fossil bodies of the older strata which have in their
general aspect, their habit or mode of growth, some similarity to
plants, and in which no organic structure can be detected beyond
sometimes the external markings."

Ferns and Cycads, but especially Ferns, have been the favourite
plants relied upon in seeking for information upon ancient climates.
Several writers lay stress upon the peculiar fitness of Ferns as tests
of climatal conditions; but Sir Joseph Hooker, whilst admitting
the possibility of drawing legitimate conclusions from their distribution in the past, mentions their wide geographical range at the
present day and their sportive nature as facts which lessen the
importance of this class of plants as a trustworthy guide.

Plants, as a whole, are considered by most of those whose
opinions have been noticed as more likely to bear the impress
of climatal changes, and therefore to afford stronger evidence,
than animals in climatological enquiries. Asa Gray[3], in a review of Saporta's book *Le monde des plantes avant l'appari-*

---

[1] Pictet (1).   [2] Solms-Laubach (2), p. 46.   [3] Asa Gray (2).

tion de l'homme, remarks—"Plants are the thermometers of the
ages, by which climatic extremes and climate in general through
long periods are best measured." In addition to the opinions on
this head already quoted we may notice a remark of Sir William
Dawson; he says[1]—"The evidence of fossil plants, when properly
studied, is, from the close relation of plants to...stations and
climates, even more valuable than that of animal fossils."

The suggestion has been made in some cases that it is possible
to recognise even among Palæozoic plants a certain connection
between genera and surface soils.

The dangers of pressing arguments from analogy too far have
been noted by many, and abundant examples drawn from the
animals and plants of the present era confirm the need of extreme caution.

The important influence which geographical position may
have exerted upon the character and distribution of ancient floras
has been referred to by Hennessy, Brongniart and others. Again,
the comparative sizes of extinct and living plants have, in numerous
instances, been made a strong argument for high temperature in
the past. That such an argument is untenable, in the case of
animals, has been well shewn by Neumayr: its application to
plants must be equally unsatisfactory except in so far as we may
reasonably consider the great size of certain fossil plants to be
proof of a climate where the period of vegetative activity was
uninterrupted by frost or darkness of long duration.

Among the earlier geologists and palæontologists we find a
general consensus of opinion that there is no indication of climatal
zones until we come to Cretaceous or even Tertiary times.

That the uniformity of the vegetation even in the Palæozoic
era, on which the conclusion as to a uniform climate has been
founded, has possibly been overestimated, we shall be in a better
position to affirm after the Carboniferous floras have been dealt
with in some detail.

That in a uniform climate, the fact that local geographical
conditions have given rise to different families or types of plants
found in the same strata, has been recognised by Godwin-Austen
and others. The danger of drawing conclusions from negative
evidence has been pointed out, and emphasized by experimental

[1] Dawson (1), (Supplementary Section, p. 1).

32   HISTORICAL SKETCH.

demonstration by Lindley: it has also been noted by Sternberg, who further warns us against placing too much confidence in Brongniart's arguments derived from a consideration of plant ratios, and that we should always remember how comparatively limited is our knowledge of the floras of the ancient world.

Since 1833, when Witham's book was published, the minute anatomy of fossil plants has been considered an important factor in deciding questions of past climates.

The degree in which annual rings are marked in fossil trees is generally regarded as throwing considerable light upon the annual variations in temperature, or upon other causes which might induce a periodicity in the activity of plant life. Witham, Corda and Hooker discussed the relative sizes of the tissue elements of some Palæozoic and recent plants, and adduced the larger size of those of the older plants as additional evidence supporting the conclusion that the climate in which those plants flourished was rather tropical than temperate.

Heer, Saporta and Neumayr have been referred to at some length as the methods which they have employed in discussing geological climates are such that, used with a reasonable amount of caution, ought to lead to fairly reliable results.

The main facts of plant distribution and plant forms and structures will be treated with more or less detail, as it is impossible to pretend to make a proper use of fossil plants as "thermometers of the past" without an acquaintance with such facts in recent Botany as bear upon the questions of station and climate.

CHAPTER II.

PLANT DISTRIBUTION.

IT is important to have a general knowledge of the present distribution of plant families, and especially of the causes which determine the several botanical regions, before venturing to apply our knowledge of plant geography in the past to the question of geologic climates. The expression "geologic climates" is a convenient one to use in the present discussion: Whitney[1] remarks "it is a term under which are commonly classed those real or supposed variations of the earth's climatic conditions, which, having taken place in past ages, before the historic period, can only be proved to have occurred by means of geological evidence." The present facts of plant distribution cannot be explained, as botanists used to suppose in the earlier period of plant geography, by regarding special floras to have been determined by climatal conditions. Climate undoubtedly plays a most important part in setting the limits to the distribution of different plants; but we have also to take into account the fact that the distribution of the present is the result of the cooperation of a number of different agents, which have been at work in past geological periods and whose actions have been to a large extent dependent on changing geographical conditions. There are several agents whose share in determining distribution is by no means unimportant. We may probably assume that the physical agents which govern the distribution of plants in our own time have acted during the several geological periods. There is, however, one important fact to keep before us in applying our knowledge of the agents of distribution now at work to plant geography in the past, and that is

[1] Whitney (1), p. 219.

the more complex nature of the present vegetation compared to that of the geological floras. As the complexity has increased through the gradual evolution and differentiation of new types, so the intensity of the struggle for existence must also have become increasingly greater. The absence in the older floras of the higher forms of plant life cancels an important factor in dealing with past distribution. It is easy to imagine how different might be the geographical range of some of the lower plants, could we remove from the field all the more highly developed forms.

The pre-Darwinian ideas caused earlier writers to attach an undue amount of weight to arguments as to climate drawn from the distribution of fossil plants, and from the proportion in which the different classes were represented, as compared with their relative position in the vegetable world at the present time.

Bentham[1] recognised three fairly ancient floras making up the vegetation of the earth. I. Northern. II. Southern. III. Tropical. The *Northern* he described as characterised by the presence of needle-leaved conifers, catkin-bearing Amentaceæ and other deciduous forest trees; many herbaceous plants, such as Ranunculaceæ, Cruciferæ, &c. This northern zone he conveniently subdivided into the following:—(a) Arctic-Alpine flora, (b) Intermediate or Temperate flora, (c) Mediterranean-Caucasian flora.

The elements of the intermediate or temperate flora are characterised by rapid growth, good means of dispersion and the capability of adapting themselves to a great variety of physical and climatological conditions. Here are characters which we may well suppose were possessed by the vegetation which spread over the same regions in Palæozoic times.

The *Southern* flora is similarly divided into zones.

The *Tropical* flora includes three subdivisions, upon whose characters we need offer no further comment.

From the point of view of the periodic phenomena in plant life depending on light, warmth and moisture, six vegetation zones have been suggested by Drude[2].

I.   Arctic, Glacial and Tundra zone.

II.   Zone of cone-bearing and foliage trees, and of summer-green moors and meadows.

---

[1] Bentham (1). Dyer (1).   [2] Drude (1), p. 83.

III. Northern zone of evergreens, with summer-green shrubs, foliage and cone-bearing trees, and of summer-hot ("Sommer-heissen") steppes and deserts.

IV. Zone of tropical evergreens, or vegetation forms whose leaf-development depends on the periodic rains.

V. Southern zone of evergreens and deciduous cone-bearing and foliage trees, of evergreen thorns and summer-dry steppes.

VI. Antarctic zone of evergreen low bushes and periodic grass and shrub vegetation.

Another grouping of plants has been instituted by de Candolle[1], in which the five divisions are determined by the biological nature of the plants, as expressed by their powers of resistance to warmth, to warmth with moisture, and to light.

I. Megathermen (or more correctly "Hydromegathermen").

II. Xerophilen.

III. Mesothermen.

IV. Mikrothermen.

V. Hekistothermen.

Drude[2] points out that we have in certain Algæ of warm springs the survivors of a once widely spread group, the Megisto-thermen. These probably agree in their biological characters with those plants which predominated in the earliest periods, and whose representatives we have in the Algæ, Ferns, Lycopodiaceæ and Equisetaceæ of the Coal-measures. The characteristics of these five groups are briefly summarized by Drude, and we are reminded of the peculiar nature of this grouping of de Candolle; his object was to establish physiological groups which would be equally applicable to past and present plant geography (*Groupes physiologiques applicables à la géographie botanique ancienne et moderne*).

We may notice the following as some of the chief agents in plant geography.

---

[1] de Candolle (1). Drude (1), p. 111.   [2] Drude (1), p. 112.

3—2

Dependence of Distribution on Geographical Conditions.

Here is one of the factors which has had a considerable share in determining our present botanical regions, and which must have had a most important influence in the earlier geological periods. Neumayr[1], in discussing the Palæozoic "Glossopteris flora" of the southern hemisphere, contrasts continental and insular floras as competing forces when spreading over new districts. He considers continental floras much better able than insular floras to hold their own in the face of competitive forms.

Engler[2] has insisted on the great importance of the power of life ("Lebensfähigkeit") as a factor in distribution, especially in the case of continental areas where the greater number of species renders the struggle for existence keener.

In the case of tropical oceanic islands we find at the present day no representatives of our continental Alpine floras, but the higher grounds are exceedingly rich in Ferns. The floras of such islands differ in a marked degree from those of continental areas, and we may reasonably suppose similar differences to have existed in the past.

The means of distribution possessed by the several families of plants are to be taken into consideration as affording a key in some cases which helps to explain facts of distribution. This applies more especially to the older floras.

As we have seen in the historical sketch it has been suggested by different writers that Carboniferous plants were peculiarly well adapted to wide distribution. The spores of the Calamitæ, Lepidodendreæ and Carboniferous Ferns would be easily scattered far and wide, either in continental areas or in an archipelago of islands. Treub[3] has furnished us with an interesting case of the rapid transference across water of seeds and spores. He visited Krakatoa three years after the eruption of 1883, and found that the flora consisted almost entirely of Ferns, only here and there a few Phanerogams near the coast or on the mountains. On the mountains of Juan Fernandez, and in Ascension Island, the flora similarly consists almost entirely of Ferns.

---

[1] Neumayr (2).　　　　[2] Engler (1).　　　　[3] Treub (1).

Grisebach[1] speaks of the sea as one of the most effectual means of preventing a mixture of different floras. This will, however, depend largely upon the means of dispersal possessed by the plants; to some the sea would not afford an impassable barrier, and would hardly prevent the spreading of such plants as the majority of those which grew in Carboniferous forests. In the case of continental islands, or in archipelagos, the sea would not form the same barrier as in oceanic islands. In recent floras it is found that when the paths along which the plants can travel are flat, and a uniform climate obtains over a large district, the floras are developed over a wide area. The Arctic vegetation and the Conifers of North Europe afford examples of such wide distribution.

Instances occur where favourable geographical conditions give rise to the distribution of species through several degrees of latitude; the absence of natural barriers counterbalancing the effect of different temperatures. Bunbury[2], in an article entitled *Some Characteristics of South American Vegetation*, mentions Desfontainea as ranging on the Andes from latitude 5° N. to 55° N. There is, he points out, a remarkable element of unity in the vegetation of South America, notwithstanding the prodigious range of temperature.

Drude[3] goes so far as to say that plants endowed with great powers of distribution could under favourable geographical conditions cover almost the whole surface of the earth.

Such facts as we have noted may well be kept before us in discussing the wide geographical range of older floras. Prof. Newberry[4], in describing a number of Rhætic plants from Honduras, has made some suggestive observations at the conclusion of his paper, which may be quoted at length.

"This discovery of a Triassic flora in Honduras is a matter of special interest, as nothing of the kind had before been met with in that section of the globe; but it is only another illustration of the uniformity of the vegetation of the world during the Triassic age. This uniformity was, however, only a development of the systematic progress of plant life. The reign of Acrogens ended with the Permian. The Rhætic epoch was therefore about the middle of the reign of Gymnosperms. Hence, after the decadence

[1] Grisebach (1).　　　[2] Bunbury (1).
[3] Drude (1), p. 100.　　[4] Newberry (1), p. 350.

of the Lepidodendra, Sigillarieæ, Calamites and Cordaites, the whole world was opened to occupation by the new dynasty of plants, the Gymnosperms (Cycads and Conifers) and the peculiar group of Mesozoic Ferns. They lost no time in entering upon their promised land and spread until they covered all portions of the, to them, habitable globe.

"Where the Gymnospermous flora originated, or how it was developed from the Acrogens, if it was so developed, and through the exercise of what elements of superiority it superseded them, we are as yet in ignorance.

"It is, however, a matter that may well excite our wonder that, migrating such immense distances from their places of origin, through every phase of soil and climate—through all the zones of the Eastern Hemisphere, and now, as we learn from this group of Honduras plants, through the New World—they marched, holding so firmly to their original group of characters, generic and specific, that wherever we open their tombs we recognize them instantly as old friends. In their long marches some perished by the way, and here and there their numbers were recruited by new forms, imported or developed; but the leading members of the troop, in virtue of some occult protection against outside influences, preserved almost without alteration all the complicated characters of their vegetative and reproductive systems.

"We shall look now with eagerness to South America for the identification there of this Mesozoic flora, which we have found in full development in Virginia, New Mexico, Sonora, and now in Honduras. It had before been recognized in Australia—where it seems to emerge from the Palæozoic flora and perhaps began— New Zealand, India, Tonquin, China, Turkestan, and various parts of Europe. Hence, with its discovery in South America we shall see it reaching as a girdle around the entire globe. This girdle was not put around the earth, however, like Puck's, in forty minutes, but in thousands and millions of years; for when we realize with what slowness the migration of plants takes place, we must recognize in the universal distribution of the Carboniferous and Mesozoic floras evidence of the lapse of intervals of time of which the duration is simply immeasurable to us."

This was written in 1888: in the same year appeared a paper

by Dr. Szajnocha[1] which seems to supply the missing link in the "girdle" spoken of by Newberry. The plants which form the subject of Szajnocha's paper are from the Province of Mendoza in the south of the Argentine Republic. Geinitz had previously described plants from the neighbouring provinces of La Rioja and San Juan and assigned to them a Rhætic age. The plants from Cacheuta in South Mendoza, as Szajnocha points out, shew a striking resemblance to plants from Tivoli in Queensland and from the Jerusalem basin in Tasmania: they agree most nearly with the Upper Trias or Rhætic of Europe.

Feistmantel[2] regards these Argentine plants as very possibly Jurassic and not of Upper Trias age; he compares them with the Wianamatta plants (New South Wales).

The height of land above sea-level and its effect on plant distribution is too well known to need a full treatment.

The occurrence of Alpine plants in widely separated latitudes, and over very wide districts, is the most striking example of the repetition of the same flora when the necessary height above sea-level is reached affording the required climatal conditions. An interesting example has been described by Johnston[3] of the difference in plants as we ascend from low to high ground, which will serve to remind us of the fact that even in a tropical climate the character of the flora alters very materially according to the height above sea-level. On the slopes of Kilima-njaro several zones of climate, as represented by the vegetation, are encountered in ascending from the Palm-covered lower ground to the snow-capped summits.

That in the floras of past periods difference in altitude gave rise to Highland or Alpine types of vegetation can hardly be doubted. Godwin-Austen and others have recognised this fact in the case of the Carboniferous flora, and, as Williamson points out, we must regard the Coniferous wood so abundant in some Carboniferous beds as the *débris* of Conifers which grew on the slopes of high ground above the level of the swamps covered with Calamites, Lepidodendra and other lowland forms. Carruthers[4], too, in

[1] Szajnocha. [2] Feistmantel (7), p. 612.
[3] Johnston, H. H. (1). [4] Carruthers (3).

speaking of the Coal-measure vegetation, expresses the opinion that the proper habitat of Conifers seems to have been higher ground.

Mr Ball[1], in an address delivered before the Royal Geographical Society, makes the extraordinary statement that it is not unlikely that in the Coal period our present Arctic-Alpine flora may have been the flora on the heights of Carboniferous mountains in low latitudes. The imperfection of our plant records is insisted on as regards the flora of mountain regions. As Mr Ball remarks, there are scarcely any cases where we have records of the mountain vegetation of the past.

Kerner[2] holds the view that in Miocene times there was an Alpine flora on the high grounds. The fossil plants of that period with which we are familiar, must be regarded as representative of the Lowland flora: they afford no data as to the Miocene Highlands' flora. We are not absolutely without proof that in Tertiary times, as at present, zones of vegetation were determined by the height above sea-level.

Saporta and Marion[3] have determined four distinct zones of vegetation in fossil plants of Pliocene age, found at different heights above sea-level in the neighbourhood of Lyons.

The lowest or littoral zone is represented by the fossils of Vacquières and l'Hérault: the vegetation, not very rich or vigorous, indicates a climate warmer and less moist than that enjoyed by the higher zones.

Populus alba, Platanus aceroides, Alnus stenophylla, Viburnum, Celastrus, Coriaria, Smilax, Acer and other plants are found associated, and the vegetation as a whole is compared to that of Syria and the African shores.

Fossils of the second zone occur in places 200 to 300 metres above sea-level, reaching as high as 600 to 700 metres. This may be called the Laurus or evergreen forest zone. Laurels, Magnolias, Anonaceæ, Ilicineæ, Tulip-trees, Liquidambars and other genera occur.

At still higher levels, from 700 to 1200 metres, are found the remnants of forest deciduous trees, Juglandeæ, Ulmaceæ, Acerineæ,

---

[1] Ball (1).   [2] Kerner von Marilaun (1).
[3] de Saporta and Marion (5).

Cupuliferæ, &c., a vegetation similar to that of the Caucasus and North America.

The fourth zone is represented by the *débris* of Conifers, recalling the vegetation of the mountains of Andalusia, North Africa and Asia Minor.

In discussing the Carboniferous flora, a possible application of this question of "Alpine" floras will be referred to in the case of the Permo-Carboniferous vegetation of the southern hemisphere.

The Nature of the Soil as a Factor in Plant Distribution.

That this is an agent which may have had its share in determining plant distribution in Palæozoic times has already been noted in reference to Sir Joseph Hooker's remarks on fossil plants[1].

Sir Charles Bunbury[2] deals with the "Influence of the chemical composition of rocks on vegetation" in an article published in his *Botanical Fragments*. He quotes from the second book of the *Georgics*, shewing that Virgil was not unfamiliar with this factor in plant distribution.

The interesting fact—noted long ago by Linnæus—is quoted by Bunbury, that saline plants thrive apparently with equal vigour in low seaside marshes and on the summits of mountains, but are seldom found in intermediate stations.

It would be superfluous to multiply examples illustrating the connection between plants and subsoils, but we may notice a few striking cases. It has been noticed that the Laterite formation of Burma[3] is characterized by deciduous as opposed to evergreen tropical woods. In many cases varieties may be connected with the nature of the soil: Asplenium serpentini, for example, is a variety of Asplenium occurring on Serpentine rocks.

Drude calls attention to the very important part played by the substratum in plant distribution. Among other examples he quotes that of the Muschelhalk plants of Germany which differ from those growing on the Bunter sandstone.

Many instructive facts, shewing the widespread action of the nature of the soil on the character of the vegetation, have been recorded by Fuchs[4]. In many instances—speaking of Italian

---

[1] Hooker (1), v. above p. 16. [2] Bunbury (1).
[3] Drude (1), p. 50. [4] Fuchs (1).

floras—he notes a change from a central European to a Mediterranean flora independent of climate. The occurrence of the Mediterranean flora he considers to depend not so much upon the climate as upon the nature of the surface rocks: it is found chiefly on calcareous rocks, whereas the central European flora replaces it on Mica-schists and clay rocks. Between Pisa and Rome a striking case occurs, shewing the influence of calcareous rocks. Up to the foot of some calcareous hills is found a central European flora, on the hills themselves a Mediterranean flora.

Grisebach has noticed that in Provence the boundary of the Mediterranean flora coincides with the limit of limestone rocks. On the shores of Lake Maggiore the same connection between the Mediterranean flora and calcareous rocks is illustrated; the sudden disappearance of Mediterranean forms resulting from the sudden dying out of the limestone.

The recognition of this important influence of the substratum explains many points in distribution, which have been considered utterly inexplicable on the assumption that climate has been the all-important power in plant distribution. In Sardinia two localities, Toveri and Tacchi, have been noticed as supporting such trees as Quercus Ilex, Ilex aquifolium and others suggestive of tropical vegetation; the rest of the island being occupied by summer-green trees. At Toveri and Tacchi the rocks are calcareous; in other parts granitic.

Kerner[1], in his *Pflanzenleben*, describes another interesting case, illustrating the same close connection between plants and rocks. The instance given, which will not be described here, is that of the vegetation in the neighbourhood of the town Kitzbühel in the north-east Tyrol, the high ground to the north and to the south being clothed with different types of plants.

Fuchs, after citing such examples as we have referred to, suggests that the want of precision by geologists in their descriptions of rocks has prevented the recognition in some cases of the connection between rocks and plants. The term crystalline schists, he remarks, is often used in too comprehensive a sense, including in one group calcareous and siliceous rocks, whose effect on vegetation has been shewn to be strikingly different. The same writer points out the importance in palæobotanical questions

[1] Kerner von Marilaun (2) Vol. II.

of such facts as he has recorded. Two floras might be found in
rocks at no great distance apart, and, having certain distinctive
characters, a different age might be assigned to each; one set of
plants, possibly, being conspicuous by their small leathery leaves, the
other by broad summer-green leaves. In such cases the floras
may be of the same geological age: we have to consider the possible effect of the nature of the strata.

The Lignite beds of Kumi are considered by Fuchs to be the
same age as the Sinigaglia beds; the difference in the fossil plants
depending on the fact that the Kumi rocks are calcareous, those
of Sinigaglia siliceous.

Starkie Gardner has suggested that the differences noticed
between the floras from the leaf-beds of Mull and Antrim may be
due to the fact that flows of basalt were continually substituting
laterite for gneiss and limestone, and this would cause a change
in the character of the forests [1].

[1] Gardner (3).

CHAPTER III.

IN attempting to picture to ourselves the conditions which obtained during the Glacial period, it is frequently forgotten that a very low temperature is not of that importance which it is often considered to be in bringing about an Ice Age.

Wallace[1] and many others have laid stress on the necessity for a concurrence of several conditions in order to render possible an abnormal extension of snow and ice.

Whitney[2], in his comprehensive monograph on *The climatic changes of later geological times*, has argued for the possibility of the former extension of snow and ice without any violent changes in climatic conditions.

After considering the question at length he remarks, " The entire body of facts presented brings out most clearly the true condition of things, namely, that the Glacial epoch was a local phenomenon, during the occurrence of which much the larger of the land-masses of the Globe remained climatologically entirely unaffected." As illustrating the possibility of glaciers existing in places whose mean temperatures differ by several degrees we may notice some observations given by Woeikof[3]. He shews clearly how other conditions than merely a low temperature are essential to ice extension. Comparing the temperatures taken at the lower ends of glaciers in East Siberia and the West of New Zealand, there is found to be a difference of more than 20°: at Irkutsk in East Siberia the temperature recorded being – 10·2°, in New Zealand 10°. Glaciers occur in the latitude of Nice and Florence extending to 212 metres above the sea-level, having at their lower ends a mean

---

[1] Wallace (1).   [2] Whitney (1), p. 387.   [3] Woeikof, A. von, (1).

annual temperature corresponding to that of Vienna and Brussels, and warmer than that of Geneva and Odessa, with a winter temperature higher than that of Florence.

Woeikof regards the theory of a general ice-covering from the Pole to latitude 45° N. during the Glacial period as absolutely untenable ; there must always have been some ice-free sea or some ice-free land, generally both.

In a recent work on *The Ice Age in North America* Wright[1] follows Ball, Whitney and other writers in shewing that too much stress has generally been laid on the low temperature necessary for widespread glacial conditions. He sums up with these words: " From all these facts it seems evident that the supposition of a slight intensification of the present conditions, so favourable to the production of glaciers in South-East Alaska, unravels the whole intricate web of glacial phenomena upon the West coast of North America."

There is the often-quoted case of the Tasman glacier descending towards the West coast of New Zealand : here the terminal face of the ice is 705 feet above sea-level, and is " hidden by a grove of Pines, Ratas, Beeches and arborescent Ferns in the foreground[2]."

All these facts shew us that the idea of the coexistence in the same region of vegetation—in some cases of an almost tropical facies—and ice-fields is not so inconceivable as one might suppose. A brief sketch of some of the conditions of plant life in Arctic lands will further make this clear, and at the same time bring out a number of facts which have an important bearing on the question of plants and climatic conditions.

Greenland.

The results of the Danish Explorations in Greenland given in the *Meddelelser om Grønland*[3] furnish us with a mass of information relating to the floras—fossil and recent—of that Arctic continent.

[1] Wright (1), p. 162.   [2] Hochstetter (1), p. 500.   [3] See p. 139.

A general account of the vegetation is given in Volume XII. of
the *Meddelelser* by Eug. Warming. He describes the plants as
occurring in different botanical regions[1].
The *Birch region* is characterised by Betula odorata, var. tortu-
osa, and Betula intermedia; with these occur a number of other
plants. It is pointed out that in Greenland, Norway and Lapland
(to the White Sea), the Birch forms the polar limit of the forests;
in Russia (from the White Sea), Siberia and America, Conifers
occupy the most northerly position. The reason for this difference
in the vegetation outposts is probably that in the Conifer districts
the air is less moist, the cold more intense and the climate more
continental. Other botanical regions are described by Warming
with their characteristic assemblages of plants. The *Rocky flora*,
consisting mostly of Mosses, Lichens and herbaceous plants, covers
a large extent of land in Greenland and other northern countries.
This flora is particularly worthy of remark as it includes a number
of species which seem proof against the severest climate, and hold
their own on the rocky peaks which rise as islands from a sea of
perpetual snow and ice. To such rocky summits above the level of the
ice-sheet Nordenskiöld has given the name "Nunatakker:" a number
of them have been explored in Greenland, and their position and
characters described in the *Meddelelser*. Prof. Kornerup[2] has been
able to collect from various Nunatakker 54 species of plants: a
list of plants is given by Warming found on Nunatakker at heights
of 4000 to 5000 feet above sea-level during the expedition of Capt.
Jensen in 1878. Even from 80° N. latitude 33 species of Phanerogams have been collected in Greenland. In Discovery Bay, 82°
42' N., and in the interior of Grinnell Land, a fairly rich vegetation has been found, rendered possible by the direct and permanent light of the sun which warms the ground and the lower
layers of the atmosphere, of "the temperature of which the mea-
surements of meteorologists, generally taken in the shade,
afford no idea." The cases of adaptation to conditions of cold
and drought, described by Warming, will be referred to later in
dealing with the general question of adaptation of plants to
external conditions.

[1] Meddelelser, vol. XII. 1888. (*Résumé* in French by Warming, p. 225.)
[2] Meddelelser, vol. I. pp. 146, 183.

Warming[1], in discussing the history of Greenland vegetation, notes the two theories as to the fate of the flora during the Glacial period.

Hooker, Buchenau, Heer and other botanists, believe that many plants survived the Ice age in Greenland; Blytt and Nathorst, on the other hand, prefer to transport the native plants to more southern and favourable latitudes during the Ice-extension, and then assume a post-glacial immigration. We need not here follow the remarks on the question of land connection between Greenland and Europe, but may notice the fact that Warming believes Greenland was not united to Europe either during or after the Glacial period.

During the maximum ice-extension many of the Greenland Highlands remained free from snow, and the country was able to keep in Nunatakker strongholds a great part of its flora. In a sketch of the work of ice-sheets, Mr Marr[2], after referring to the Danish explorations in Greenland remarks, "At the time when the Pennine Chain was nearly buried by the great ice-sheet from the west and north-west, the parts which stood out above the ice may still have possessed a meagre flora, for plants are found upon Jensen's Nunatakker which are situated about forty English miles from the edge of the ice, and have an elevation of over 5000 feet."

Mr Clement Reid[3], in a recent number of *Nature*, dissents from Warming's view as to the Greenland vegetation during the ice age. He agrees with Nathorst, and considers it would have been impossible for any Phanerogams to have survived in the few non-Glaciatal "Nunataks;" they would be too high, and the lowland all snow and ice.

Clement Reid compares the condition of Britain after the deposition of the boulder clay with that of Greenland at the present time: of this Greenland type of vegetation we have relics in such plants as Salix polaris, S. reticulata, Betula nana, found in Suffolk and a few other localities.

This question as to the possibility of the survival of the Greenland flora during the Ice age has recently been discussed at some length and with no little energy by Nathorst[4] and Warming[5].

[1] Meddelelser, xii.   [2] Marr (1), p. 5.   [3] Reid (3).
[4] Nathorst (1) (4).   [5] Warming (1).

According to the latter an essential part ("Kern") of the Greenland flora survived, on ice-free islands of rock, the rigour of the Glacial period. Nathorst quotes Steenstrup and Kornerup as to the former extension of the ice-sheet in Disco island and other places: ice scratches and erratics forcibly suggest that ice at one time covered Jensen's Nunatakker. Considering the greater mass of ice, and the increase in snowfall which must have accompanied the Ice age, Nathorst concludes that probably any peak above the ice would be covered with snow, and most likely all the Phanerogamic vegetation succumbed to the extreme Arctic conditions. Even if bare peaks stood out above the ice sheet it by no means follows that their rocky surfaces would support vegetation.

Warming replies in detail to Nathorst's arguments, and shews that several species of Phanerogams occur in Greenland at altitudes higher than that of the peak to which Nathorst refers, as an example of a mountain in part free from snow and entirely without Phanerogams. There are, moreover, other conditions to be considered than height above sea-level and temperature, in connection with the presence or absence of plants. In cases where the exposed rocks are readily disintegrated the abundance of loose detritus might prove an effectual obstacle to Phanerogamic plants. The arguments brought forward on both sides as to whether Greenland was connected by land with Europe in post-glacial times need not be discussed here.

Grinnell Land.

Lieut. Greely[1] has given a list of plants collected in the neighbourhood of Fort Conger in latitude 81° 44'.

Alaska.

From the accounts of recent explorations of the coast of Alaska we have further proof that vegetation is able to hold its own under conditions, which might not unreasonably have

---

[1] Greely (1), vol. II. App. ix.

been considered fatal. Drude[1], endeavouring to shew the error of associating Arctic conditions with an absence of vegetation, refers to Seton-Karr's expedition to Alaska, which affords a striking confirmation of the fact that plant life exists, even in abundance, in the midst of ice and snow. Seton-Karr describes the Alpine regions of Alaska as furnishing the most extensive and thickest glaciers after Greenland and the Arctic regions. Some of the glaciers coming down from Mt. Elias, which rises almost sheer from the ocean's edge to a height of 20,000 feet, are covered at their ends with brushwood growing amongst the loose *débris* lying on the ice.

Not only does vegetation spread over the moraines of the retreating glaciers but even—in favourable summer temperatures—covers the moraine on the ice surface. The snow-line of Mt. Elias is a little above 100 metres: the mean temperature on the coast in January is − 8°C., in July 14°C.

The following extracts from Seton-Karr's book describe this association of ice and vegetation[2].

" Further progress directly north towards Mt. St Elias became barred by a huge buried glacier, overtopped by immense masses of moraine and overgrown thickly with shrubs and Fir trees, which were becoming disordered and destroyed when they grew on the edge or faces of the moraines."

In describing the great Agassiz glaciers Seton-Karr writes[3], " Under these piles of moving stones, which are for ever being carried forward, lies the glacier ice, three or four hundred feet in thickness at the edge and much thicker elsewhere; while a tangled forest of Spruce and Birch, Maple and Alder, is growing along its extremity, so thickly and closely, that it becomes exceedingly difficult, especially to men with large packs on their backs, to force a way through."

H. W. Topham[4], in the *Alpine Journal* for 1889, gives a similar account of Mt. Elias and its glaciers; he speaks of a dense vegetation growing upon the moraine-covered ice.

In a paper contributed by Drude to *Petermann's Geographische Mittheilungen* for 1889, we find the opinion expressed—based on Seton-Karr's account of Alaskan glaciers—that portions of the

---

[1] Drude (2).   [2] Karr, Seton (1), p. 77. See also Karr, Seton (2).
[3] Karr, Seton (1), p. 85.   [4] Topham (1).

Scandinavian forest vegetation probably lived in sheltered spots in
the midst of the north European Ice-sheet. Nathorst[1] has dealt at
some length with Drude's remarks, and dissents from the view that
the northern forest vegetation could have survived the Ice age.
The case of Alaskan glaciers is very different to that of a continent
clothed in a sheet of ice: the comparison must be made between
Scandinavia in the Glacial period and Greenland to-day: in Alaska
we have only glaciers, and hence the conditions are not strictly
comparable with those which obtain in Greenland.

Nathorst goes on to say that probably during the Glacial epoch
no peaks stood out as Nunatakker above the surface of the Scandinavian Ice-sheet. Possibly this inland ice of Scandinavia was
bordered here and there by plants other than true Arctic species.
The evidence afforded by recent "finds" of fossil plants shews that
for a long time after the melting of the Ice-sheet the climate was
so severe that only extreme Arctic species such as Salix polaris,
Dryas octopetala and others were able to exist. According to
Nathorst, the fossil plants shew that an Arctic flora fringed the
border of the inland ice in southern Sweden and Denmark:
Drude's theory of the existence of forest trees receives, in fact, no
support from such fossil Arctic plants whose remains have been
found. The Cromer beds are quoted by Nathorst, as shewing that
previous to the Ice-sheet Arctic plants lived in Norfolk; the trees
found in the "Forest bed" are relics of a forest vegetation killed
off by the severe climatal conditions.

No doubt as Nunatakker appeared above the sinking Ice-sheet
an Arctic flora would obtain a footing on the rocky peaks, and
later, when moraine matter, freed by the melting ice, accumu-
lated on the surface, plants might spread as on the moraines of
Alaskan glaciers.

Fossil Arctic plants will be briefly referred to later under
Pleistocene Botany; such testimony as they give lends support to
Nathorst's views and makes it difficult for us to subscribe to
Drude's opinion that even at the height of the Arctic tide of ice
forest trees might still find shelter and nourishment in high
northern latitudes.

¹ Nathorst (4).

Russian Lapland.

In the *plant biological studies* of Kihlman[1] we have a mass of material brought together, relating not only to Lapland but to Arctic regions generally. He gives a number of instances of early growth in Arctic plants, quoting also observations of Kjellman and Nathorst. In his remarks on the limits of vegetation in the Kola peninsula, he states that it is the climate, and not any change in the ground, which sets a limit to the woods. One of the points which Kihlman brings forward prominently, is the effect of wind upon plants and the part it plays in determining vegetation limits. It is not simply the mechanical force of the wind, not the cold, nor yet the amount of salt or moisture in the atmosphere, which forms an impassable barrier to trees, but chiefly the drought, which lasts for months and dries up the young shoots at a time of year when no compensation for evaporation of water is possible. In Lapland the same adaptation of plants to external conditions is met with as in Greenland and other countries.

From this short and imperfect sketch of some of the more noteworthy features of the vegetation of high northern latitudes we have gained some insight into the characters of Arctic floras, and the extremes of temperature which they are capable of withstanding.

It is instructive to note not only the adaptations of plants in Arctic regions to the severe climatal conditions, and their great power of resistance to low temperatures, but also to notice what are the effects of abnormally severe frosts upon plants in more southern latitudes. A number of observations have been recorded by Lindley[2], Göppert[3] and others bearing upon the effect of severe frost on plants. In questions of this kind one must distinguish between freezing and freezing to death (" Gefrieren " and " Erfrieren "). The temperature at which death from freezing ensues varies considerably, depending upon the specific constitution of the plant's protoplasm.

Nitella syncarpa, which is frozen to death at $-4°C.$, and Sphærella nivalis, which lives at a temperature of $-20°C.$, serve

---

[1] Kihlman (1).　　[2] Lindley (1).　　[3] Göppert (2).

to shew the wide limits within which the fatal temperature may
fall. In the Alps we have many familiar instances of the association of plants with snow-covered ground. Kerner von Marilaun
in his *Pflanzenleben*[1] gives a coloured plate shewing the Alpine
Soldanella melting its way through a stratum of "Firn" at an
altitude of 2240 metres on a Tyrol mountain.

The protecting powers of snow—how it prevents radiation from
the earth and also protects plants from cold winds—are well
known, and instances are numerous enough of its power of shielding plants from freezing to death. Drude[2] furnishes us with
striking illustrations of the capability of plants to withstand severe
cold. In northern Siberia, for example, at Werchojansk on the Yana
in latitude $67\frac{1}{2}°$ N. Siberian Larches grow; the mean January
temperature is $-49°$ C., with a minimum of $-60°$ C. and a maximum of $-28°$ C. We are ignorant of the real causes of death in
plants by freezing. It is a well-known fact, however, that very
much depends upon the state of development of the tissues when
they are subjected to these severe temperatures.

It is an interesting fact that Alpine plants are often frozen if
cultivated in the warmer plains: they prefer the long winter rest
of Alpine heights and the regularly recurring spring, to the constantly changing climate of the plains. Drude speaks of the protection afforded by snow to Polar vegetation, but notes the fact that in
many northern regions where there is no permanent winter snow
there exists nevertheless an Arctic flora in the summer. We have
abundant proof that perennial plants are able to remain frozen for
a considerable part of the year without being killed. After the
retreat of a glacier in Chamounix[3], several plants—Trifolium
alpinum, Geum montanum &c.—were found to continue growing
after being covered four years by an advancing glacier. Göppert
cites many cases shewing the effect of frost upon vegetation. He
remarks that the vegetation of Polar lands is an expression of the
warmth, which the surface layer of the earth and the lowest air
strata receive from the summer sun.

He mentions a case of some fruit trees which, after being sent
to Russia, lay twenty-one months frozen in a cellar and afterwards
revived. There is another striking case quoted from Middendorf

[1] Kerner (2), vol. i.   Plate opposite p. 465.
[2] Drude (1), p. 24.   [3] Pfeffer (1), p. 435.

who described a Rhododendron parviflorum in Siberia which had
its younger branches in flower whilst the upper parts of the stem
and root were frozen stiff.

In the case of Algæ we have the well-known instance, recorded
by Kjellman[1], of certain genera of marine Algæ on the coast of
Spitzbergen enjoying vigorous life, and carrying on the processes of
reproduction, at a temperature of $-1°$ C. during the three months'
Polar night.

The facts and generalisations which have been quoted are of
great importance as serving to prevent us assigning too high a
temperature to those geological periods of whose vegetation fragments have been found in Polar latitudes. It is true we have
indisputable evidence, as the late Prof. Heer has so well shewn,
for a warmer temperature in past times than now obtains within
the Arctic circle; but when we remember how plants are able
to effectually stand against even the most severe climatal conditions, and allowing a more favourable distribution of land and sea
before the Glacial period, we may yet be able to explain the facts
of climatal changes in Arctic regions without falling back on the
theory of the shifting of the earth's axis. To such a theory
physicists refuse to give their sanction; and, moreover, the circumpolar distribution of fossil plants affords a serious obstacle.

In a recent contribution to Glacial Geology by Mr G. W.
Bulman[2], several instances are given to illustrate the possibility of
vegetation flourishing in close proximity to glaciers. In addition
to such cases as have been already noted, there is one quoted by
Bulman, where a temperate vegetation occurs side by side with
glaciers, from Rendu's *Glaciers of Savoy* in which the Glacier
des Bossons is graphically described advancing between banks
covered with flowers and near enough to a field of rye for
the ears to be blown against the glacier ice. In the Himalayas
the close association of ice and vegetation is described by Colonel
Tanner in the Sat valley; "here may be seen forests, fields,
orchards, and inhabited houses all scattered about near the ice
heaps[3]."

The facts of Arctic vegetation, as I have already pointed out,
must help us in questions connected with the fate of plant-life

---

[1] Kjellman (1).   [2] Bulman (1).
[3] Bulman (1), p. 407.

during the maximum development of the Great Ice age. We have abundant proof that we cannot entirely accept the opinions of those who would have us believe in a general retreat and partial extinction of plants during the advance of the northern Ice fields: there can be little doubt as Forbes[1], Warming, Drude and others have maintained, that the injurious effects on vegetation of wide-spread Glacial conditions have been greatly overrated.

[1] Forbes (1).

CHAPTER IV.

THE INFLUENCE OF EXTERNAL CONDITIONS UPON THE MACROSCOPIC AND MICROSCOPIC STRUCTURES OF PLANTS.

Habit and Size of Plants in relation to Climate.

THE inference has often been drawn, in the case of Coal-measure plants particularly, that the woody structure and tree-like form of plants, whose descendants we are familiar with as genera without woody stems, point to tropical conditions of climate.

Size has been considered by many an argument in favour of tropical climates.

To refer briefly to a few points bearing on this question.

Plants with woody stems are able to live through the winter of the cold temperate zones, because the lignification of part of the plant tissues is followed by a development of cork, a screen against cold.

In some cases a protection is also afforded against cold by the collenchymatous nature of the subepidermal tissues; these cells with their thickened angles, characteristic of collenchyma, affording a shield against cold to the more delicate tissue of the phloem and cambium. Areschoug[1] has illustrated this in the case of Leycesteria formosa.

In the warm temperate zone woody plants increase in number, and often substitute an evergreen for a deciduous habit. In the tropics, too, woody plants are abundant. There the wood is not a safeguard against the influence of cold, but serves to give the plants that firmness which they require to enable them to support their branches. In a tropical climate, cork must be looked upon, not as a screen from cold, but as a

---

[1] Areschoug (1).

regulator of transpiration of which it prevents excess. Thus we find that the same contrivances, and the same structures, may serve several ends. Areschoug points out, that in such questions as he deals with in his paper on *Der Einfluss des Klimas auf die Organisation der Pflanzen*, we have to remember that not only climate, but also other external conditions under which a plant lives, find expression to some extent in its structure and form; climate, however, plays the most important part in its influence on plant organisation.

Hildebrand[1] has given some instructive generalisations which have reference to the influence of external conditions on plant structures.

There are, he remarks, two kinds of climates; a uniform climate, and a climate with periodic changes. A uniform climate favours the existence of perennial and woody plants. It must not be forgotten that a uniformity of climate does not exclude an annually recurring period of rest, a pause in vegetative activity, not necessarily the result of cold or other unfavourable conditions. Hildebrand gives instances of uniform climates, and the comparative abundance, in places with a slightly varying temperature, of woody plants belonging to families or genera which have no woody representatives in latitudes where a periodic climate obtains.

In the West Indies plants with woody stems are abundantly represented. In the Sandwich Isles, where the climate has no periodic interruptions, woody plants occur in families which in periodic climates have no woody genera. Isodendron (Violaceæ), Alsinodendron (Caryophyllaceæ), Geranium arboreum, and three tree-like Compositæ are examples.

Similar instances might be quoted from St Helena, Madeira, and the Canary Isles.

A periodically changing climate, on the other hand, favours short-lived plants and shrubs instead of woody trees.

If we consider plants according to their distribution, and notice such characters as are typical of different regions, we find that certain floras are distinctly characterised by well-marked peculiarities in the forms and structures of their members. In a paper of exceptional interest by Tschirch[2], a number of plant

1 Hildebrand (1).   2 Tschirch (1).

types are described and classified under several zones which are determined, to a large extent, by the distribution of rainfall. A similar grouping of "natural types" is recognised by Grisebach[1] and others. The contrast between floras of dry and wet regions is one of the most striking facts brought out by such investigations as those of Tschirch. This contrast is expressed, not only in the form and structure of the leaves, to which we shall pay special attention, but also in the habits and general morphology of the plants.

Succulent plants have been described by various writers as characteristic of dry climates; a general account of this class of vegetation has recently been given by Goebel in his *Pflanzenbiologische Schilderungen*[2]. In districts where water is scarce the succulent tissues are essential for storage of water. Such plants as Cacti, and other dry-climate forms, are provided with long roots[3], which penetrate below the dry upper layers of the soil to moister lower strata, from which they absorb water to be conducted into the various reservoirs of the plant. Reserve supplies of water may be stored up in succulent stems, branches, leaves or tubers.

As examples of plants whose swollen and succulent stems are conspicuous, we may refer to Goebel, who quotes Martius' account of the Catinga floras of Brazil, and other instances.

In Steppe regions of the Old World succulent Chenopodiaceæ are especially abundant, taking the place of the American Cacti.

In Cacti[4], and similarly constituted plants, various anatomical peculiarities have been described, which are specially concerned in storing or economising water. The highly cuticularised outer walls of epidermal cells, the thick hypoderm, the structure, number and arrangement of stomata are some of the adaptations to a dry climate well illustrated by Cacti; to say nothing of the reduction of leaf-surface and the handing over of leaf functions to specially modified stems and branches. Euphorbiaceæ, Crassulaceæ and many other families adapt themselves to scarcity of water, or intense illumination—and

[1] Grisebach (1).   [2] Goebel (1).
[3] Good examples of the long roots of such plants were seen during the Suez Canal excavations. Volkens (2).
[4] Caspari (1).

therefore excessive evaporation—by developing succulent tissues such as we find in Cacti and Chenopodiaceæ.

Succulents must not be regarded in all cases as necessarily indicative of steppes or arid districts; they are represented in sea-coast floras by species of several families which grow in salty places, and protect themselves against the evil effects of strong salt solutions by the development of such characters as are typical of Xerophiles. Examples of succulent species of Caryophyllaceæ, Cruciferæ, Umbelliferæ &c., are given by Goebel from sea-coast vegetation.

Among tropical Orchids the succulent type is represented by certain species as described by Krüger[1]. Thus we find a recurrence of the same type of structure in plants growing under very varied conditions.

Another and very important characteristic of floras exposed to intense sunlight and drought is the *reduction of evaporating surfaces*.

Such a reduction may take various forms, and be developed to a greater or less extent according to circumstances. Examples of this means of protection and adaptation are sufficiently well known, and do not call for more than a passing notice of the numerous types.

We may have leaves entirely suppressed, or represented by small scales and spines, and the reduction of surface by stems functioning as stems and leaves combined. Phylloclades, or leaf-like stems, suggest such conditions of climate as render essential a reduction of ordinary leaf-surface. Phyllodes, or flattened and leafy petioles, are examples of similar conditions, and, in certain floras, e.g. in Australia, play a very conspicuous and important part. Tschirch[2] has pointed out the abundance of the slender Acacian Phyllodes, and other instances of leaf reduction, in the Australian flora.

Not only are there various genera, species and even families of plants which can be recognised as members of desert, steppe and dry climate floras, but occasionally xerophile structures are met with in plants usually regarded as characteristic of moist habitats. In addition to such cases as have been previously mentioned we may note that of Dicksonia antarctica[2], which is able to withstand a

---

[1] Krüger (1).   [2] Tschirch (1).

certain amount of drought by reduction of the intercellular spaces and extra cuticularisation of the epidermal cells of its leaves.

We see, then, that a dry region may be reflected in the habits and structures of its floral elements; and more localised conditions of drought make themselves apparent by the appearance, in some members of a flora, of structural details known to be typical of dry conditions. The bearing of such facts as these upon the question of ancient climates is sufficiently obvious.

It has, however, been shewn by various botanists, that we cannot always refer corresponding structures or characters of plants to the same kind of external conditions. The resemblances between typical Xerophiles and sea-coast plants have already been hinted at: a full account of such is given by Schimper[1] in his sketch of the Java floras. Schimper demonstrates by several examples that protections against transpiration are adopted by plants under varied conditions, and for dissimilar reasons: similar adaptation being found in typical Xerophiles, in Halophytes, in Java Alpine plants, and in evergreen woody plants of colder temperate zones. The xerophile characters of the Alpine flora of Java are expressed in the structures and habits of the plants, and instances are referred to from other countries shewing the recurrence under different conditions of similar adaptations of structure.

Leaves, their form, position and structure.

Seeing to what an extent the palæobotanist has to content himself with detached leaves of plants, it would be extremely helpful if he were able to depend upon their shape, venation or internal structure as guides in settling the question as to the climatic condition under which the plants flourished.

The cases in which fossil leaves are found with internal structure preserved are exceedingly rare. One may consider, however, how far we are able to connect anatomical structure with external conditions of growth in the case of living plants.

After describing the mechanical principles of leaf venation, Sachs[2] remarks, "These considerations, however, teach us at the same time how fruitful is every conception of organic forms based

[1] Schimper, A. F. W. (1).   [2] Sachs (2), p. 53.

on the principle of causality, as compared with the merely formal comparison of the same. Ettingshausen has, in the latter sense, described thousands of leaf-venations, without arriving at any important results whatever, while our simple principle, according to which the leaf-venation fulfils, on the one hand, the carrying to and fro of nutritive matter, and, on the other, the mechanical office of keeping the lamina tightly extended and protecting it from rupture, affords a very clear insight into the variety of forms here prevalent."

Tschirch[1] points out that if a plant type is the expression of a certain kind of climatic peculiarity, this must impress upon it some anatomical character, visible in the size of leaf, the consistence of tissues, or in the form and method of arrangement of leaves. The precise connection between the characters of leaves and external conditions is not as yet fully understood.

The form of leaf, or extent of surface exposed to the sun, is naturally dependent to a certain extent upon atmospheric conditions, and serves to regulate transpiration.

In moist places, where there is no danger of too rapid transpiration, the leaf-surface may attain a considerable size : in places where the atmosphere is dry, and too great a loss of water has to be guarded against, the leaf-surface is reduced. In desert plants, for example, the leaf-surface is small : this is well shewn in Spartium, Genista and other Leguminosæ. This reduction of leaf-surface under such conditions is further seen on comparing desert and European species of Genista.

The trees forming a wood in temperate latitudes—say in South Germany—are characterised as a whole by their leaves being flat and without hairs. A difference may be noticed in the size of leaf according as the locality is shady or sunny, the same species having its leaves larger when growing in shady than when growing in sunny places[2]. Convallaria polygonatum illustrates this ; growing in shady places its leaves are at least three times as large as when exposed in sunny places. The Beech affords another example illustrating variation in the size of leaves. Stahl[3] points out that Beech leaves increase gradually in size as the situation becomes more and more shady. Certain Alpine plants

---

[1] Tschirch (1).   [2] Kerner (2).   [3] Stahl (1).

seem to afford exceptions to this rule : the leaves of those growing
at high altitudes and exposed to the sun's rays are not less than
the leaves of those growing in shady places at lower levels. This
finds an explanation, however, in the abundance of fogs and consequent moisture in the atmosphere on the high ground.

The reduction of evaporating or transpiring surfaces of plants
in dry climates is especially noticeable, either in the entire absence
of leaf laminæ or in the decrease of their size. Areschoug[1] has
instanced Rubus Australis, which serves to shew how a plant may
alter the extent of its leaf-surface according to circumstances; in
shady places the leaves of this species have feebly developed
laminæ, in exposed situations only a midrib and petiole. Sorauer[2],
Scott-Elliot[3] and others have shewn that the same plants in a
moist atmosphere have larger leaves than in a dry atmosphere.
Spiny and leathery leaves are characteristic of dry climate plants,
but, like other xerophile features, are represented in various floras.

The vertical position of leaves is worth noting as an additional
test of dry climates : Tschirch has drawn attention to this character in Australian plants.

Increase in the size of leaves favours transpiration, and is
necessarily regulated by such external conditions as render transpiration rapid or slow. The immense leaves of Corypha umbra-
culifera are good examples of the connection between the extent of
leaf-surface and atmospheric conditions unfavourable to rapid
transpiration.

Rolled leaves. The rolled form of leaf characteristic of steppe
grasses and desert plants is sufficiently well known without calling
for detailed description[4]. Kerner von Marilaun, in enumerating
the various contrivances by which leaves insure the maximum
amount of transpiration in a given time, includes the rolled form
of leaf and considers the chief purpose of such leaves is to keep
the stomata free from water, thus enabling transpiration to go
on freely.

Kihlman[5], after describing rolled leaves in Russian Lapland,
refers to Kerner's remarks, and insists upon the connection between
this type of leaf and conditions of drought; the main object of the

[1] Areschoug (1).   [2] Sorauer (1).   [3] Scott-Elliot (1).
[4] Tschirch (2).   [5] Kihlman (1).

rolled form being to expose as little of the stomata-bearing surfaces as possible to evaporation.

Warming[1], in his description of the vegetation of the rocks and dry plains of Greenland, refers to the rolled leaves as special adaptations for protection against drought. The Greenland form of Juniperus communis has smaller leaves, and more closely applied to the stem, than the ordinary form of the species: the same fact is noticed in Lycopodium selago and L. annotinum. The recurved edges of the leaf lamina, e.g. in Empetrum nigrum, tend to reduce transpiration. There is also an abundance, in the Greenland flora, of the "Pinoid type" of leaf, another form by which surface reduction is effected: the nature of the epidermis and the small number of the stomata both serve to reduce evaporation.

Warming calls attention also to the shrubs of rocky districts whose leaves are smaller than those of the same species growing in other countries.

This rolled character is conspicuous also in the Venezuela "Paramos" flora as described by Goebel[2]. In this flora xerophile characteristics are abundant: Goebel mentions the following as checks to transpiration adopted by "Paramos" Compositæ:—

(i)   Thick felt of woolly hairs.

(ii) Development of leathery leaves.

(iii) Rolled leaves.

(iv) Reduction in size of leaves; and other characters of less importance in the present discussion.

In these Venezuelan "Paramos" there is no sharply defined period of drought and no necessity, so far as the presence of water is concerned, for hindrances to evaporation. But in spite of the presence of water in sufficient abundance there is another cause, namely temperature, which necessitates water economisation. Transpiration goes on vigorously, aided by wind and a dry atmosphere; but in the ground at certain seasons the temperature is too low to allow of a sufficient supply of water being absorbed by the roots: hence xerophile characters appear in spite of moist surroundings.

<hr>

[1] Meddelelser, vol. XII. pp. 107—116, figs. 1—14.

[2] Goebel (1).

The dry "Punas" of Peru, referred to by Goebel, offer a marked contrast, so far as climate is concerned, to the "Paramos" of Venezuela, but their floras possess many points of resemblance.

The minute structure of Leaves and its relation to external conditions.

We have seen to what an extent the form or size of a leaf may be influenced by temperature, drought or other circumstances.

There is still the question of the form and arrangement of the tissues enclosed between the two epidermal layers of a leaf. How far is the form of the assimilating tissue of a leaf an expression of outside conditions?

In the *Botanische Zeitung* for 1880 Stahl[1] gives the results of his investigations on the structures of leaves and how far they are influenced by conditions of light. His previous researches in connection with chlorophyll had led him to suspect that the degree of intensity of certain physiological phenomena would be expressed in details of structure.

He found that leaves exposed to strong light have a greater development of the pallisade form of assimilating tissue than leaves in shade. The thallus of Marchantia polymorpha and several leaves of Phanerogams are given as examples of the effect of strong or feeble light on the structure of assimilating tissues.

Haberlandt[2] notices the variation in the form of the assimilating cells of leaves under different conditions, but considers that light, as the most important external factor of assimilation, has scarcely any influence on the arrangement of the assimilating systems.

Similar results to those arrived at by Stahl were obtained by Pick[3], who found that development of the pallisade form of tissue depends upon the intensity of light to which the leaves are exposed. In the main, Pick agrees with Stahl, but prefers to regard the pallisade cells in leaves growing in shady places as merely shortened pallisade cells, and not a distinct form of assimilating tissue. It is possible, he says, to recognise in the

---

[1] Stahl (2).   [2] Haberlandt (1), (2).   [3] Pick (1).

modified tissues of leaves growing in the shade, the same number of cell layers as occur, in the normal pallisade form, in leaves exposed to a stronger light.

Figures are given which shew clearly the effect of variation in light upon the leaf tissues in such plants as Hedera Helix, Hieracium villosum &c.

In 1883 another communication appears from Stahl[1], which describes more fully than his earlier paper the modifications of leaf tissue. He notes that a spongy form of leaf parenchyma is especially strongly developed in plants growing in a moist climate. In places where the climatal conditions render excessive transpiration disadvantageous, pallisade parenchyma is more conspicuous.

It is found that such leaves as the Beech illustrate clearly variations in size and structure depending upon conditions of illumination.

Beech leaves in sunny places are sometimes three times as thick as the leaves of the same species exposed to feebler sunlight : in the sunny leaves nearly all the assimilating tissue takes the pallisade form ; there is also, as one would expect, a difference in the cuticle in the two cases.

Leaves of other plants shew in an equal degree this difference in structure expressing variation in light intensity. Figures are given by Stahl[2] of transverse sections of Beech leaves, Lactuca scariola, Ilex aquifolium, &c.

The hypoderm tissue expresses also the effect of sun and shade. This is especially noticeable in the leaves of Conifers : in Abies alba, for example, the leaves growing in sunlight have an almost uninterrupted layer of hypoderm below the epidermis ; in shady situations this hypoderm layer is but poorly represented.

In considering the influence of climate on the cuticularisation and lignification of Coniferous leaves, Noack[3] remarks that the needle-like leaves become more lignified, the higher the position of the trees above sea-level, and the farther they are North or South of the Equator.

The remarkable stiffness of the leaves of dry climate plants is a well-known fact, and is due to the existence of ribs or groups

[1] Stahl (1).   [2] Stahl (1), Plate x.   [3] Noack (1).

of strengthening elements which give elasticity to the tissues. Tschirch[1], in speaking of such leaves, points out the importance of distinguishing between structures whose function is merely mechanical and those which serve some physiological purpose. It is impossible to enter fully, in this brief sketch, into the various microscopic details of leaf structure which characterise plants in which it is of importance to prevent excessive loss of water by transpiration. We will merely enumerate some of those anatomical features which might be recognised in well-preserved fossil plants, and which would afford important evidence in questions of geological climates.

Epidermal outgrowths.

Plants possess in hairs another protection against too rapid transpiration. Gnaphalium leontopodium (" Edelweiss ") affords a well-known example of an Alpine plant thus protected against the risk of being dried up in exposed situations with little soil. The leaves of many plants are found to be hairy or glabrous according to the latitudes in which they occur. Some plants which range from North to South Europe, from Scandinavia to the coast of the Mediterranean, are characterised by smooth leaves in the North, and in the South are provided with a protecting covering of air-filled hairs. As examples Kerner quotes Silene inflata, Campanula speculum, and other plants whose leaves in North and Central Europe are smooth and glabrous, whereas the same species in Calabria have hairy leaves.

Pick and others have also drawn attention to the fact that an abundance of hair-like outgrowths affords a protection against excessive evaporation in dry sunny places.

Vesque and Viet[2] found by experiment, that an increase in the number of hairs on hairy leaves is brought about by growing such plants in dry situations. Hairs, again, may be regarded as shielding plants against too rapid temperature variations. Plants growing in open districts, where the changes of temperature are great, have more hairs than those growing in woods where the temperature is more even, moisture more abundant and evaporation less.

[1] Tschirch (1).
[2] Vesque and Viet. *Ann. Sci. Nat.*, vol. XII., 1881, p. 167.

Hairs when full of air, and thickly spread over the leaf-surface, act as a sheet of felt and hinder evaporation; also, as Volkens [1] suggests, during the night the "felt" plays an important part in absorbing dew for Desert plants. The hairy nature of plants other than true Xerophiles has already been noticed.

The same hairy and fleshy nature of the plants is conspicuous in coast species of the Algerian flora, and in species from the highest peaks of the Atlas.

The Size and Number of Intercellular Spaces.

Another hindrance to evaporation is effected by a limitation of the intercellular spaces in leaf tissues. Tschirch notes Eucalyptus sp. as an example: species in valleys or on river banks have many intercellular spaces; species growing in the "Scrub" have compact and thick-walled assimilating tissue [2].

In this particular, too, coast vegetation resembles that of dry climates; as shewn, for example, by Schimper in his description of the Mangrove flora of Java.

Stomata.

The position and form of the stomata are among the most useful tests of the conditions under which a plant grows. This has been made clear by the work of Tschirch [3] and other botanists. Tschirch uses the types of stomata as a basis of classification.

It has been proved by experiment that certain plants placed in dry situations reduce the amount of transpiration by a greater cuticularisation, and also by reducing the number of stomata. Without attempting to enumerate instances which have been recorded shewing the close connection between stomata and external conditions, we may notice some examples quoted by Kerner.

Kerner [4] deals at length with the position of stomata in cavities or grooves, and considers the sunken position favourable to transpiration. Not being exposed directly on the leaf's surface, but separated from it by a small ante-chamber or groove, there is a better chance of free passages being preserved for the exit of

[1] Volkens (2).
[2] Additional examples are given by Stahl and others.
[3] Tschirch (1), (2).   [4] Kerner (2), vol. I. p. 266 *et seq.*

aqueous vapour from the intercellular spaces of the plant. Hakea florida, Protea mellifera, Dryandra floribunda and other plants are mentioned as examples. In moist periods, when it is of the utmost importance to the plant that transpiration may go on as actively as possible, the sunken stomata are of great service in keeping open a free passage. When the moist periods are succeeded by periods of drought, reduced transpiration being then essential, this position of the stomata affords a protection against too active evaporation.

Kihlman[1] considers Kerner's statements somewhat contradictory, and emphasizes the fact that the primary function of stomata, not directly on the leaf-surface, is to protect the leaves against such rapid evaporation as would result in drying up the plant tissues.

By comparing species of Eucalyptus and Grevillea taken from dry and moist localities, Tschirch found that not only were the "dry" species characterised by sunken stomata, or stomata in grooves, and by thick cuticles, whilst species from moist places had their epidermal cells less cuticularised and their stomata on a level with the leaf-surface, but there was also noticed a decrease in the number of stomata as the conditions of drought became more pronounced.

Sunken stomata, and stomata in hairy grooves, are, also found in the Java Mangrove vegetation. In Conifers and Cycads we have stomata below the level of the leaf surface, but in these cases clearly not expressions of drought. Tschirch suggests that probably structural differences between the Angiosperm and Gymnosperm types of stomata would hardly admit of a direct comparison between Cycads and Conifers on the one hand, and Angiosperms on the other as regards the connection between stomata and external conditions[2].

Cuticularisation.

This is one of the means of protection against the drying up of plants, instances of which have been incidentally mentioned. A thick and uniform cuticle characterises evergreen leaves, but the two do not invariably occur together: in Caryota propinqua, for example, the cuticle is not strongly developed, and the leaves of

---

[1] Kihlman (1).
[2] Volkens (1) has also dealt with this subject at some length.

this Palm are not protected against drought. Plants whose leathery and stiff leaves are provided with a highly developed cuticle are generally considered to be adapted to dry conditions. Whether the dry and leathery nature is also a protection against cold, seems, according to Kihlman, to be doubtful, as we find the number of evergreens increases in districts where the winters are mild and the periods of drought of long duration. In Russian Lapland, Kihlman noticed many leathery-leaved plants growing in moist places: in such cases the leathery nature is, he considers, a protection against the winds; even Bog plants, with their moist substratum, are in danger of being dried up when exposed to the Polar winds.

Covering of Wax.

The waxy covering of epidermal cells serves to exclude water in the form of rain or dew, and further protects the leaf against the loss of certain substances which would readily diffuse out from surfaces in immediate contact with water. Plants with wax coverings to their leaves are described by Warming from Greenland, and by Kihlman from the Kola Peninsula, and in many other cases which need not be quoted here, as this is a character hardly likely to be of much value in dealing with fossil plants.

Water-Plants.

In considering the connection between plant forms and structures and conditions of life, we must not entirely omit a reference to water-plants. Although not strictly within the limits of the present subject, the chief characteristics of plants living wholly or in part in water, may be briefly noted. Among Coal-measure plants, whose structures are more or less faithfully preserved, there are some which have been compared by palæobotanists to plants living in swamps or submerged in water. In cases such as these it is of primary importance to have clearly before us the most constant and characteristic differences between land and water plants, and especially such as are likely to be detected in the mineralised tissues of fossil plants.

The work of Dr H. Schenck[1] clearly demonstrates that water-plants, taken as a whole, constitute a well-defined and natural

[1] Schenck, H. (1), (2).

group, both as regards external form and anatomical structure. In submerged plants the leaves are in nearly all cases finely divided or ribbon-shaped; their delicate structure is expressed in the relatively fewer layers of parenchymatous cells as compared with the leaves of land-plants, and mechanical tissue is very feebly developed. The substitution of diffused for direct sunlight results in the disappearance of pallisade tissue. Submerged leaves are further conspicuous by their more delicate epidermal cells whose outer walls have a very thin cuticle, by the absence of stomata and of that dorsiventral structure so common in land-plants: moreover, in water-plants there is no need for any of the various checks to transpiration, which have been previously discussed [1].

In the leaves and stems of submerged plants the presence of large and small intercellular spaces and canals is characteristic. In water-plants with *floating leaves* we find reniform and large oval assimilating surfaces, stomata on the exposed epidermis, pallisade tissue present, and indeed a much nearer approach to the leaves of land-plants than in the case of submerged leaves.

Without giving a complete account of the forms and structures of water-plant leaves, we have mentioned a sufficient number of facts to indicate the existence of well-marked peculiarities, rendered necessary by the medium in which the plants live and by the consequent differences in the process of feeding.

In the stems of water-plants we may note the following characteristics: very little mechanical tissue, great reduction in the number and differentiation of the xylem elements, more delicate nature of the bundle-sheath, absence of secondary thickening, presence of lysigenous or schizogenous air spaces, general absence of secretory and excretory receptacles, &c.

As regards the roots: many submerged plants are without these organs, and where such are present they are but poorly developed, but their bundle-sheaths are more differentiated and suberised than in stems.

In a more recent paper Schenck[2] has called attention to the development of "aerenchyma" in water-plants; a tissue homologous to cork in land-plants, and consisting of cells whose walls are not suberised and which are separated by numerous intercellular spaces.

[1] See also de Bary (1) p. 31.   [2] Schenck, H. (3).

Acclimatisation and Naturalisation.

Having thus passed in review a few of the better known facts illustrating a close connection between anatomical structure and external conditions, we may note a few cases of acclimatisation and naturalisation of plants.

In discussing the subject of acclimatisation, Drude refers to the possibility of cultivating tropical plants in glass-houses in spite of the absence of tropical sunshine.

Northern plants transported to Madeira grow actively and keep up their period of rest, which in more Northern latitudes occurs in winter, by shedding their leaves and remaining leafless a certain length of time each year. In the case of Alpine plants it is found that they may be kept in cold houses over the winter, and, although their specific minimum temperatures are not reached, they pass through a period of repose at a much higher temperature than in their native places.

Plants from distant lands often become naturalised in our own latitudes where they have to contend against native species already accustomed to the climate.

Engler[1], in speaking of Eocene plants, quotes Martius, who refers to certain Southern plants being capable of withstanding a very low temperature in the gardens of Montpelier.

Many plants, it is found, which are included in the tropical class, are not injured by exposure to a cold winter.

Instances might be given of plants where nearly related species are able to live under very different conditions of climate.

In some observations made by Naudin[2] on plants in the gardens of Collioure we have interesting facts recorded which bear on the present question. In the winter of 1870—71 the temperature was abnormally low, and frost lasted much longer than usual. A number of plants are mentioned which fell victims to the severity of the winter, whilst several were able to tide over the unfavourable conditions without suffering any permanent harm.

Agave Americana, Mesembrianthemum edule, Livistomia Australis and others suffered considerably but were not killed. Chamærops humilis, Jubæa spectabilis, &c., survived. After discussing a number of examples shewing the behaviour of tropical

¹ Engler (1).   ² Naudin (1).

plants in temperate latitudes, M. Naudin concludes : " The unequal resistance of plants to climatic influences, and the geographical distribution which results from this, are amongst the most obscure problems of vegetable physiology; and we have no means of explaining why in the case of plants very closely allied morphologically, when placed under the same physical conditions, some are destroyed by the cold and others suffer no harm."

For an explanation of such facts as these, one must look to some other cause than a difference in the character of the plant tissues.

Johnston[1], in his *Kilima-njaro Expedition*, speaks of certain species of plants on the flanks of Kilima-njaro, whose generic home is in the hot tropical plains, but which have strayed up the mountain slopes and accustomed themselves to the colder conditions. Others again from temperate regions have ventured down the mountain, and have become used to higher temperatures.

We see from the above short notes on the questions of adaptation and acclimatisation of plants, that we must exercise great caution in speculating as to climatic conditions of geological periods from such knowledge of the tissues of fossil plants as their fragmentary remains have afforded us.

Minute Anatomy of Fossil Plants.

In living plants it is possible to use certain characteristics of internal structure as indices of external conditions. We have seen how clearly marked, in many cases, is this influence of climatal conditions upon the structure and form of leaves.

As our knowledge of the internal structure of fossil plants increases we may hope to learn something from microscopic investigations as to the climates of past ages.

Such specimens of fossil leaves as have been examined, with their minute structure preserved, do not afford any striking anatomical contrast to the tissues of living leaves.

In the Permian and Carboniferous rocks numerous examples of Cordaites have been described in which the minute anatomy

---

[1] Johnston, H. H. (1).

is almost perfectly preserved. Such are figured by Renault[1], Grand'Eury[2], Felix[3], and others.

A surface view of the leaves has revealed sometimes well marked and numerous stomata, apparently on a level with the cuticle, and not sunk below the surface of the leaf. The parenchyma of the leaves seems to have been fairly compact or dense, but intercellular spaces were by no means entirely absent. Sclerenchymatous or strengthening tissue is abundantly represented in Cordaites leaves. Cordaites, we may note, does not agree exactly with any living genus or even family: its structure and general habit have been thoroughly investigated, and the researches of Grand'Eury and Renault on this plant form a striking instance of the minuteness and thoroughness of microscopical examinations of fossil plants which are rendered possible by silicification of their tissues.

Renault refers Cordaites to a distinct family (Cordaiteæ), which, in its male flowers, agrees with Salisburieæ, in the female flowers with Cycads, and in the structure of the wood with Conifers.

Solms-Laubach[4], after describing Cordaites leaves, remarks: "That with all this the structure of the leaves is essentially resistant has been already remarked by Schenk; we see by this case the great antiquity of these anatomical phenomena of adaptation to external conditions; we shall find them reappearing on different occasions as we proceed with our subject, and we may conclude on the whole that those external conditions, which we see determine this adaptation at the present day, prevailed as early as the period of the Coal-measures."

What light does the structure of Cordaites leaves throw upon the question of Permo-Carboniferous climate?

Broadly speaking we see no indication that these leaves were exposed to any conditions of climate other than such as now obtain. Stahl[5], in his paper on leaf structures in living plants, to which reference has already been made, suggests that the discovery in the Coal-measures of a single leaf shewing strongly-developed pallisade parenchyma would justify the inference that there was not that thick atmosphere, which is so frequently spoken of, as

[1] Renault (1), Pl. xvi.   Renault (2), Pl. xxii.   [2] Grand'Eury (1), Pl. xviii.
[3] Felix (1), Pl. iii.   [4] Solms-Laubach (2), p. 107.   [5] Stahl (1).

everywhere shading the Palæozoic forests, but, in places at least, there were gleams of sunlight, the existence of which would be registered in the form and arrangement of the cells of the leaf tissue. Solms-Laubach[1] cites the bifacial leaves of Cordaites, with pallisade cells on the upper side, as affording evidence such as Stahl speaks of.

It is not uncommon to find in coal, fragments of cuticular structures or cell-membranes: the most notable example of this is the "paper coal" ("Blätterkohle" or "Papierkohle") of Tula (Russia); this is made up in places of strips of cuticle from Bothrodendron punctatum (Lepidodendron tenerrimum). These cuticle flakes, so remarkably preserved in the Malöwka coal beds, have been described by Göppert, and later by Zeiller[2]. The cuticularised cell walls, more resistant than other tissues, have not fallen victims to Bacillus amylobacter or other destructive agencies.

In the English Coal-measures leaves and leaf-bases of Lepidodendra are not uncommon; their internal structure is fairly homogeneous, and shews no special characteristics which call for further notice.

Fern pinnules also have been figured by Williamson[3] with their tissues more or less intact; but the English specimens fall far short of those from Autun, which have been described and figured by Renault: as examples of the latter Pecopteris geriensis and Alethopteris may be mentioned. Here we find pallisade parenchyma well developed and suggestive of direct sunlight. In certain pinnules Renault[4] has detected what he considers to be waterglands, similar to glands found in some living Ferns.

Schenk[5] has shewn the importance of examining microscopically the epidermal layers of fossil leaves: such an examination in many cases supplies us with important aids in the determination of genera. In the case of Mesozoic Conifers[6] the microscopic examination of epidermal cells has more than once been the means of correcting determinations based upon mere external resemblances. Pinus Crameri (Heer) illustrates this; the form of the epidermis cells and the structure of the cuticle shewed that Cyclopitys, and not Pinus, was the genus.

[1] Solms-Laubach (1).   [2] Zeiller (1).
[3] Williamson (2). *Phil. Trans.* 1874, Pl. LII.   [4] Renault (2).
[5] Schenk (1) (2).   [6] Schenk (3), p. 293.

No great help seems to be afforded by such methods of examination, at present at all events, in questions of geological climates.
Solms-Laubach[1] describes and figures a leaf of Ullmannia
selaginoides in which the sunken stomata are fairly well preserved, situated somewhat below the level of the leaf surface.
Pallisade cells occur all round in Ullmannia leaves where the
structure is radial, and not bifacial as in Cordaites.

In his account of the fossil flora of the Keuper and Lias
Schenk figures specimens of Baiera, Sagenopteris and Thinnfeldia
shewing epidermal cells and stomata.

Zeiller[2] follows Schenk in making use of epidermal structures
in deciding questions of plant determinations; he gives figures
of Cycadopteris Brauniana and Frenelopsis Hoheneggeri: in the
former the stomata are situated in a groove, comparable to the
stomata in Cycas revoluta; in the latter the stomata are very
numerous and have a peculiar feature in the four- or five-rayed
arrangement of their orifices. A somewhat similar form of stomata
is found in Cupressineæ, and also in Marchantia.

Instances are given by Conwentz[3] of Conifer leaves (Pinus),
with epidermal cells and stomata partially preserved, but nothing
of importance is noticed as to the forms or positions of the
stomata.

- We cannot as yet learn many lessons in Climatology from
the structure of stems, roots or other parts of fossil plants. Attention has been called by various writers to the comparatively feeble
development of wood in Carboniferous Vascular Cryptogams, in
proportion to the size of their stems; also to the supposed succulent nature of the cortical tissue. In the stems of Lepidodendron
and some other Upper Carboniferous plants there was a considerable development of certain radially arranged elements in the
cortex; this structure has been compared to the periderm of
some recent plants.

For examples of cortical periderm structures we may refer
to Williamson's figures[4].

The outer layers of this cork-like tissue separate readily from
the more internal tissues, and thus we frequently find hollow

---

[1] Solms-Laubach (1).   [2] Zeiller (1).   [3] Conwentz (1).
[4] Williamson (2).   Phil. Trans. 1872, Pl. xxiv. xxv.   Phil. Trans. 1878, Pl.
xx., &c.

shells of Lepidodendroid stems which have been squeezed together, or filled with sand or clay. Solms-Laubach[1], in speaking of Grand'Eury's theory of the formation of coal, writes as follows in explanation of this common method of preservation:—"To explain these peculiar conditions Grand'Eury supposes that the temporary raising of the level of the water in the basins of accumulation flooded the flat swampy forest-ground far and wide, and that the trees were killed by the inundation and became rotten and at last fell to pieces, their stumps only remaining erect beneath the water. Such behaviour is quite conceivable, if we take into account the small development of wood and the succulent nature of the rind in the trees of the Carboniferous period; and that something of the kind does take place in warm climates I was able to satisfy myself in the Botanic Garden at Buitenzorg, where a colossal Palm-tree, which had died after developing its terminal inflorescence, broke up and fell in pieces before my eyes with a startling crash."

The increase in thickness which characterised the Lepidodendreæ and Calamitæ of the Coal-measures and, according to Prof. Williamson's recent account of Lyginodendron, possibly occurred in Ferns, is a striking feature of Carboniferous vegetation, and one which forms the most important line of demarcation between living genera and their Palæozoic ancestors.

To what cause must we ascribe this? Luxuriant growth is generally taken to be characteristic of tropical conditions, and so indeed it is: we have previously pointed out the fact that certain families of Phanerogams have a larger percentage of woody genera in warm countries than in the colder climate of Europe. The inference, however, is hardly warranted from such facts as these that because the Coal-measure trees had this distinguishing characteristic of secondary increase in thickness, therefore the climate of that time was much warmer than the present climate of Europe. We have several facts to take into account in questions of this kind.

Considering the wide gap which separates Coal-measure plants from their modern descendants, we cannot safely argue from analogy as regards habit or conditions of growth. Again, during

---

[1] Solms-Laubach (2), p. 24.

the Carboniferous period these Vascular Cryptogams were at the
zenith of their power, they spread over a wide area with the
vigour characteristic of plants in their prime, and unopposed by
such competitors as have arisen since the Palæozoic era.

Since Schwendener's[1] great work appeared on *Das mechanische Princip im Anatomischen Bau der Monocotyledonen* much
light has been thrown upon the significance of mechanical
strengthening tissues in plants. Schwendener notes, for example,
in the case of Psaronius that the roots had their cortical tissue
protected by a peripheral sheath of Sclerenchymatous elements; this
genus of Tree-ferns, growing in moist places and with numerous
air canals in its cortex, would require protection against pressure.

In the gigantic Lepidodendra (60—100 feet high), the same
botanist points out, there must have been great thickness and
strength in the more external layers of the cortex, or a considerable
thickness of the woody axis.   Vigorous development, expressed in
secondary thickening of stems and roots, does not, we venture to
think, justify the conclusion that the climate of the Carboniferous
period was more tropical in these latitudes when hundreds of
square miles were covered with Palæozoic forests than it is to-day.

In a description of a specimen of Caulopteris punctata from
the Upper Greensand of Shaftesbury in Dorsetshire, Carruthers[2]
draws attention to the existence of five slight constrictions in the
stem, and considers it probable that they afford indications of such
seasonal interruptions of growth as are characteristic of temperate
latitudes. This inference is based upon the fact that similar
constrictions have been detected in stems of British Ferns, but
not in tropical or sub-tropical Ferns.

In the case of more recent fossil trees, Godwin-Austen[3] and
James Geikie[4] have called attention to the great thickness of the
bark in some stems found in peat bogs: this is considered proof
of colder conditions; the tough resinous wood and thick bark of
our bog Pine, says Geikie, bear emphatic testimony to the rigour
of the seasons.

[1] Schwendener (1).   [2] Carruthers (4).   [3] Godwin-Austen (1).
[4] Geikie, J. (1).

CHAPTER V.

"ANNUAL RINGS" IN RECENT AND FOSSIL PLANTS.

THE presence or absence of annual rings in fossil wood has been quoted as bearing directly upon the question of climates. Witham [1] was one of the first to notice the difference between Palæozoic, Mesozoic, and recent woods, as regards the degree of distinctness of rings of growth. Unger [2] expresses the opinion that it is of importance for us to be familiar with the conditions which govern the formation of rings in living trees, in order that we may be the better able to speculate upon the significance of their presence or absence in fossil trees. Annual rings, says Unger, may be used as indicators ("Fingerzeig") with regard to climatal conditions. In the Palæozoic Coniferous woods he describes the annual rings as entirely absent or very indefinite: they become much better marked in Keuper and Liassic Conifers, in the Oolitic woods still clearer, and in Tertiary times the wood of Conifers appears to have closely resembled that of recent trees. It is concluded from these facts that in the older periods there was no marked change in temperature during the year, not even such variations as now occur in the tropics. The term "annual ring" has come to be generally applied to the layer of tissues added every year by the activity of the cambium to the wood of a tree. In tropical countries, these annual additions are not marked off from one another in the same degree as in the wood of trees whose period of vegetation is interrupted by a season of cold or wet, as in extra-tropical countries. In the great majority of trees growing in climates where there is an annually recurring period of suspension in the activity of growth, the annual rings are clearly defined. During each season of growth there is a sharp

---

[1] Witham (1).   [2] Unger (1).

separation between wood formed in spring and wood added in the autumn; that part of the wood formed in the early part of the year has been called *Spring wood*, that formed in the latter part of the year *Autumn wood*.

Strasburger[1] has suggested the term *late wood* (*Spätholz*) in preference to autumn wood. He quotes in support of this new term the fact—pointed out by Th. Hartig—that trees complete the addition to their wood about the end of August. So far as the wood is concerned the formation of new elements is confined to two-and-a-half months of the year.

To enquire: (i) What constitutes the lines of separation, between the different elements of wood, which appear as annual rings?

(ii) Do the limits of a so-called annual ring always represent the wood formation of one year?

(iii) How is the formation of "annual rings" explained?

(i) Sachs[2] has described the main points in the structure of wood rings, and, as we shall see later, suggested an explanation of their formation. As a rule the rings are easily seen with the naked eye, and also in microscopic sections. In some cases a cursory glance at a transverse section of wood may reveal what appear to be annual rings, when such are not really present.

Schleiden[3] describes a transverse section of some wood taken from the lid of a Chinese box: in this case a number of dark lines were noticed running at right angles to the medullary rays and suggestive at first sight of annual rings; further examination proved these to be prosenchymatous bands which did not form regular circles, but occupied slightly different positions on opposite sides of the medullary rays.

Fischer[4] has shewn that occasionally coloured bands occur in autumn or spring wood, which may be mentioned as a possible source of error in calculating the ages of trees.

Binney[5] has pointed out that in some fossil woods differences in colour have been mistaken for annual rings.

The line of demarcation between early and late wood may be due to one or more of the following causes.

[1] Strasburger (1), p. 500.   [2] Sachs (1).   [3] Schleiden (1).
[4] Fischer (1).   [5] Binney (1).

(a) A shortening in a radial direction, and flattening in a tangential direction, of the elements at the outer limits of the late wood

(b) Very often the late wood elements have thicker walls than those of the spring wood.

(c) The distribution of non-equivalent forms of tissue in the year's wood.

In some plants there is no increase in thickness of the walls of the late wood elements (e.g. Alnus, Fagus, &c.). In Pinus sylvestris, where the rings are well marked, the autumn tracheids have a diameter a quarter the size of the spring tracheids, and their walls twice as thick.

Cases are given by Russow where annual rings are either absent or indistinctly marked.

Sanio[1] speaks of trees without annual rings as very rare and only mentions Muhlenbechia complexa. De Bary[2] refers to floating woody plants as characterised by wood of uniform structure, and points out that the wood of the White Nile Ambatsch (Herminiera) shews no distinct rings.

In cases where no tangential flattening occurs, the separation of the annual additions is much less distinct. The wood of Cytisus is quoted by van Tieghem[3] as an example. In some species of Araucaria there is a tendency towards suppression of the annual rings. In Araucaria excelsa the rings are clearly seen with the naked eye, but with difficulty detected under the microscope ; there is only a narrow zone of thin-walled spring elements. There are various other plants which need not be quoted in detail in which a suppression of annual rings is noticed.

Sachs[4], in speaking of annual rings, remarks :—" In tropical woody plants when several years old, the additions to the wood formed in each successive year are not generally distinguishable on a transverse or longitudinal section ; the entire mass of wood is homogeneous." Unger states that rings occur in tropical trees, but never clear and sharp.

The walls of the elements vary in thickness, but the diameters remain constant throughout the year.

[1] Sanio (1).   [2] de Bary (1).   [3] van Tieghem (2).
[4] Sachs (1), p. 132.

(ii)   In the *Botanische Zeitung* for 1844[1], a case is given shewing the formation of two rings each year.   Perottet, during his stay in Senegal, saw a Baobab tree felled in which he counted sixty concentric rings: natives, who saw the tree planted, gave its age as 34 or 35 years.

During a year two periods of drought are experienced in Senegal which cause interruption in the activity of growth, and apparently give rise to two rings each year.   Another case is mentioned by Unger: in 1846 the summer was unusually long and had a marked effect upon the growth of trees; this was shewn in the development in woody plants, not only of the buds of that year, but also of those destined for the following season.   The result of this abnormal activity was found to be expressed in the structure of the wood, where two rings were formed instead of one; the boundary between the two rings formed during the year was found to differ from that between the normal rings.   Unger compares the structure of the wood added in that year to the wood of tropical trees.

Kny[2] gives a list of references to double rings and describes a case which came under his observation where two rings were formed in one year.   The leaves of Tilia parvifolia were eaten by the caterpillars of Lymantria dispar about the end of June, and this caused the premature development of the next year's buds.   The branches on which these buds had unfolded shewed two rings instead of one.

Strasburger[3] notes the doubling of annual rings in the case of Larix.

Another instance is given by de Bary[4]: Adansonia digitata forms two rings annually, the result of two periods of vegetation falling in one annual period.

Conwentz[5], in his excellent monograph on Baltic Amber trees, refers to the occurrence of double rings in fossil and recent Conifers.

Other references, in addition to those I have already alluded to in speaking of the occurrence in living trees of double "annual" rings, are given by Göppert[6] in his account of Conifers from the Tertiary Amber beds.

---

[1] *Bot. Zeitung*, "Kurze Notizen," p. 367 (1844).   [2] Kny (1).
[3] Strasburger (1), p. 24.   [4] de Bary (1).   [5] Conwentz (1).
[6] Göppert and Menge (3).

Göppert points out how the character of the annual rings varies in the same species, and depends upon the nature of the ground, height above sea-level, latitude and other causes.

(iii) Sachs[1] has given a simple explanation of the formation of annual rings, which has generally been quoted in text-books as affording a satisfactory explanation of the differences between vernal and autumn wood. As it is essential for us to be familiar not only with the structural characteristics of annual rings, but also with the possible cause of their formation, it will be well to enquire into the theories which have been advanced to account for their existence. Sachs attributes the difference in size of the wood elements, which is so sharply defined in many instances, to a difference in the amount of pressure under which they are formed in the spring and autumn. In spring the pressure exerted by the bark on the cambium is less than in the later part of the year; as wood formation goes on the pressure of the cortex gradually increases, and the elements resulting from cambium divisions become gradually smaller.

Sachs considered his hypothesis confirmed by the experiments undertaken by de Vries [2].

De Vries' experiments only dealt with such cases where the annual limit is marked by a flattening of the elements of the late wood. Without describing his experiments we may notice the results arrived at. Under artificially increased pressure it was found that the autumn wood was formed at a time when, under normal conditions, spring wood would have been added. When pressure was relieved by cutting into the bark the cambium gave rise to wide elements at a time when late wood ought to have been formed.

De Vries concluded:—(i) The number of cell divisions in the cambium are dependent upon pressure.

(ii) The growth of elements in radial and tangential directions is dependent upon pressure.

(iii) The proportion between wood fibres and vessels depends upon pressure. The greater the pressure the fewer vessels are formed.

Doubt was thrown upon this pressure theory by Russow, Kny, Krabbe[3] and others. If pressure is the cause of this well-defined difference in size of the wood tissues, why do not the cortical elements shew a similar expression of pressure variation ?

Again, according to this theory, the wood of tropical trees should not shew any signs of spring wood and late wood; but as a matter of fact annual rings are not altogether absent in tropical trees.

Russow suggested the presence in a greater or less quantity in the developing elements of a substance which readily takes up water, causing thereby a greater or less turgescence.

Krabbe undertook experiments with a view to determine what variations, if any, occurred in the pressure exerted by the cortex on the cambium. He was led to believe that, during the formation of an annual ring, no perceptible difference is manifested in the intensity of the cortical pressure.

Sachs' theory, with the apparent confirmation of de Vries' investigations, was considered by many unsatisfactory, and shewn by experiments to be founded upon incorrect data.

According to Krabbe, de Vries did not take into account the pathological results of his experiments.

If Russow's suggestion as to the variation in turgidity be correct, there ought to be a perceptible difference in the hydrostatic pressure in spring and later in the year. Wieler[1] made a number of observations to determine if such were the case, and found that the hydrostatic pressure at the time of formation of the late wood was not less than in spring.

He found that differences in turgidity did not occur at any definite season. Wieler came to the conclusion, that although Russow's turgidity theory could not be maintained, he still agreed with him that difference in nourishment is the primary cause of the variation in size of the early and late elements of the woody tissue. Late wood, according to Russow and Wieler, is the result of poor nourishment.

Hartig, on the other hand, considered late wood to be the result of better nourishment.

Wieler, without considering the question of annual rings as settled, comes to the conclusion that experiments have proved them to be entirely dependent upon conditions of nutrition.

---

[1] Wieler (1).

Strasburger does not regard either the pressure or nutrition theory applicable to certain cases.

We see from the above sketch that botanists are still in the dark as regards the real cause and significance of the formation of rings of growth. That certain differences in nutrition, and variations in the stimuli acting on the cambium cells, find expression in the variation in size or number of the wood elements, seems to be established. It has been shewn that the explanation of the formation of annual rings is not quite so straightforward and simple as is often supposed. We have not sufficient data at present with regard to rings of growth in trees of different latitudes, nor an adequate knowledge of the cause of variation in the cambium products, to allow us to speak with any great weight upon questions of past climates, so far as evidence is afforded by the examination of transverse sections of fossil trees.

The fact that there is no winter rest in tropical countries, does not preclude the possibility of an interruption to the progress of vegetation resulting from some other cause than winter cold. Many tropical plants flower all the year round, and there is no interruption to the thickening of their stems: others pass through a resting period brought on by long-continued drought.

These facts are of importance as warning us against assuming a necessary connection between tropical climates and a homogeneous structure in the stems of woody plants. On the other hand it is unsafe to connect a periodic cessation of active growth with an annually recurring fall of temperature, or some other seasonal interruption to the vegetation period. The facts given by Heer as to certain trees remaining leafless for a definite period every year in Madeira, where the temperature is fairly uniform, afford examples of this.

We must take into account the necessity of rest to enable plants to elaborate food material previous to the unfolding of new leaves. We may conclude by quoting Sachs'[1] remarks on the question of dormant periods. "This periodic alternation of vegetative activity and rest, is in general so regulated that for a given species of plant both occur at definite times of the year, leading to the inference that the periodicity only depends upon

[1] Sachs (2), p. 350.

the alternation of the seasons, and therefore chiefly upon that
of temperature and moisture. Without wishing to deny the
co-operation of these factors, a closer consideration shews, however, that this matter must depend chiefly upon changes which
take place in the resting plant, independently of external influences, or only indirectly affected by them."

Rings of growth in fossil plants.

We have seen that in living trees the presence or absence, or
size of annual rings, cannot be relied upon with certainty in the
determination of genera, bearing in mind to what variations the
structure of the wood of trees is subject, in the same family or
genus. In describing fossil Coniferous wood, Schenk[1] reminds us of
the absence of annual rings in some Araucariae, and occasionally
in Gingko.

Those who have paid most attention to the structure of the
wood of Conifers, agree that annual rings can only be considered
as confirmatory evidence in questions of determination. Stenzel[2]
remarks upon the variable character of annual rings both in stems
and roots of Conifers, and considers them of little importance from
a taxonomic point of view.

The arrangement of the elements in annual rings has been
shewn by Mohl to be different in roots and stems, and the difference, being a constant one, may be of service in dealing with fossil
woods.

It may be noted too, that the annual rings in Cupressaceæ,
Podocarpeæ and Taxeæ are generally smaller than in Abietaceæ.
There are certain general characteristics of this kind, which are
not without value when used with other characters; and this
known variation in the rings of living Conifers serves to shew how
little weight should be attached to any conclusions based upon
annual rings in dealing with fossil forms. It is at all events
certain that in living woods the absence of annual rings is
distinctly exceptional. In examining the structure of fossil woods,
especially those from Palæozoic strata, do we find that it is
exceptional for annual rings to be present? This question is often
answered in the affirmative, and deductions drawn accordingly
about absence of seasonal changes and the like at the time when
these trees flourished.

---

[1] Schenk (3).   [2] Stenzel (1).

We cannot here give a detailed account of fossil woods, nor indeed, in many cases, have sufficient observations been made with a view to determine whether annual rings are present or absent; but a few cases will be enumerated which will sufficiently prove that concentric rings of growth are by no means universally absent from Palæozoic plants.

Silurian.

In Nematophycus (Prototaxites of Dawson) microscopic sections have revealed the existence of distinct concentric rings of growth in the tissues of the stem. These rings are the result of a variation in the diameter of the tubes making up the mass of the plant. Dawson[1], in describing Protaxites, mentions the concentric rings of growth. Carruthers gives a figure of part of a transverse section of Nematophycus which shews clearly one of the rings. Dawson's determination has been shewn by Carruthers to be erroneous, and botanists now generally recognise in this Silurian and Devonian genus a gigantic Alga. Prof. Penhallow[2], in a description published in 1889, notices the rings but considers them to be independent of the cause which determines rings in exogenous plants. Mr Barber, who has in preparation a description of some new specimens of Nematophycus, informs me that "annual" rings are distinctly marked: he compares them with similar concentric rings in some species of Laminaria and Alaria[3]. Whether or no we may connect this periodically recurring variation in the activity of growth with the annual shedding of a "leaf," such as takes place in Laminaria digitata, it is certain that the structure of Nematophycus indicates some regularly recurring alteration in the conditions of growth. We see at all events that even in a marine Alga regular rings of growth may be detected.

Devonian and " Sub-Carboniferous."

Mr F. H. Knowlton[4] has recently published a list of species of Cordaites, Dadoxylon, and other fossil woods from different formations : he clears up much of the confusion which has resulted from the use of different systems of nomenclature. Clear definitions are given of the genera Cordaites, Araucarioxylon, Dadoxylon, &c. For convenience I have noted the presence or absence of annual

[1] Dawson (1).   [2] Penhallow (1).
[3] Hooker, Pl. 167 and 168, *Figure of Lessonia sp.*, (2).   [4] Knowlton (3).

rings in three columns: in many of the descriptions quoted by Knowlton nothing is said about concentric rings of growth; it does not of course follow that in all these cases no rings were present. Doubtless imperfect preservation, in many cases, has prevented the recognition of that difference in the size of the walls, or cavities of the tissue elements, to which annual rings are due.

	Rings absent, or not mentioned	Rings indistinct	Rings distinctly marked
Cordaites Ouangondianus (*Daws.*).			+
C. Halli (*Daws.*).	+		
C. Newberryi (*Daws.*).			+
C. Clarkii (*Daws.*).	+		
Dadoxylon Beinertianum *Endl.*			+
D. Tchichatcheffianum *Endl.*			+
D. Buchianum *Endl.*		+	
D. Vogesiacum *Ung.*			+
Araucarites Ungeri *Göpp.*		+	

Araucarites Ungeri is not mentioned by Knowlton: Stenzel speaks of annual rings expressed in the tangential flattening of zones of the wood elements. The beds from which this species is recorded, Stenzel[1] notes, are classed by Zimmermann with the Culm strata.

Carboniferous.

	Rings absent, or not mentioned	Rings indistinct	Rings distinct
Cordaites Brandlingii (*Lindl. & Hutt.*).	+		
C. intermedius (*Grand'Eury*).	+		
C. Stephanense (*Grand'Eury*).	+		
C. Subrhodeanum (*Grand'Eury*).	+		
?C. Acadianum (*Daws.*).		+	
C. materiarum (*Daws.*).		+	
Dadoxylon ambiguum *Endl.*			+
D. angustum (*Felix*).			+
D. annulatum *Daws.*		+	
D. antiquum (*Witham.*).		+	
D. medullare *Endl.*	+		
D. antiquius *Daws.*			+
Araucarites Thannensis *Göpp.*	+		
A. carbonaceus *Göpp.*			+
Pinites Conwentzianus *Göpp.*			+
Araucarites Elberfeldensis *Göpp.*	+		

Prof. Williamson[1], in his Memoir VIII, figures a transverse section of Dadoxylon (Cordaites) shewing concentric rings of growth. So far we have referred to Conifers and Cordaites: we may, however, look to other Coal-measure plants, and see what evidence they afford as to annual rings.

Dawes[2] was the first to record annual rings in Calamites: he speaks of them as being comparable to the rings in stems of recent Dicotyledons as regards the clearness of their definition.

Binney[3], in his description of sections of a Calamodendron (Calamites) stem, mentions pseudo-annual rings, due to successive deposits of mineral matter in the elements of the wood. Seeing that there are no certain traces of concentric rings of growth in the transverse sections of Calamites figured by Williamson, Stur[4], and others, we may not unreasonably suspect that Dawes was mistaken, and possibly was led astray by such deceptive appearances as are spoken of by Binney and Stenzel.

No annual rings can be detected in Stigmaria, Lepidodendron, Sigillaria, Asterophyllites, &c.: at least so far as Prof. Williamson's figured specimens are concerned. It is, however, not impossible that more minute investigation, undertaken with the special object of determining annual rings, may bring to light indications of them not hitherto noticed. In some of the transverse sections of Stigmaria figured by Williamson[5] there are, however, indications of zones of interrupted growth.

In a transverse section of Amyelon radicans, figured by Williamson[6], rings of growth are clearly shewn having a peculiar excentric arrangement. Amyelon is regarded by Williamson as probably the root of Asterophyllites. Felix[7] also describes the rings of growth in the same species, but considers Amyelon to be the root of a Conifer. Solms-Laubach[8] agrees with Felix as to the systematic position of Amyelon, and calls attention to the

---

[1] Williamson (2) *Phil. Trans.* 1877, p. 230, Pl. IX. fig. 46.
[2] Dawes (1).   [3] Binney (1).
[4] In the specimen figured (from Stur) by Solms-Laubach there are three distinct concentric rings, but these are probably due to other causes than periodic variations in vegetative activity (Solms-Laubach (2), p. 298).
[5] Williamson (1), Pl. VIII.   [6] Williamson (2), *Phil. Trans.* 1874, Pl. IX. fig. 56.
[7] Felix (1).   [8] Solms-Laubach (2), p. 364.

necessity of further investigating the ringed structure of the wood, since, as de Solms remarks, "a true formation of yearly rings has never been observed in any other remains from the Carboniferous formation[1]."

Permian.

	Rings absent, or not mentioned	Rings indistinct	Rings distinct
Cordaites medullosus *Göpp.*	+		
Dadoxylon Schrollianum (*Göpp.*).			+
D. stigmolithos *Endl.*	+		
D. stellare *Ung.*		+	
D. Valdejolense (*Moug.*).			+
D. Rollei *Ung.*		+	
D. Richteri *Ung.*		+	
D. Saxonicum (*Geinitz*).			+
D. pachyticum (*Göpp.*).			+
D. Rhodeanum (*Göpp.*).			+
D. Fleurotii (*Moug.*).			+
D. Permicum (*Merckl.*).		+	
D. cupreum (*Göpp.*).		+	
D. biarmicus (*Kutorga*).			+
D. Keuperianum *Endl.*	+		
Tylodendron (Araucarioxylon) *Weiss.*	+		
Araucarites Permicus *Merckl.*			+

Tylodendron. The real nature of this genus of Weiss has recently been recognised by Potonié[2]; he considers it undoubtedly a true Araucarioxylon, in the narrower sense of the name, that is a true Araucarian[3].

Araucarites Permicus. This, like Tylodendron, is not mentioned by Knowlton. Under Dadoxylon Permicum *Merckl.* he gives Araucarites Permicus *Merckl.* as a synonym, but Araucarites Permicus described and figured by Ludwig[4] is not mentioned: it is worthy of note that Ludwig describes the annual rings as strongly marked.

[1] This hardly agrees with some of the descriptions of Coniferous wood by Stenzel and Knowlton. Since the above was written I have obtained specimens of Coniferous (?) wood from a sandstone of Coal-measure age in Yorkshire, which shew annual rings with remarkable distinctness.

[2] Potonié (1).

[3] If this conclusion of Potonié is correct we must carry back Araucaria from the Mesozoic to the Palæozoic formation. The term Araucarioxylon, as suggested by Felix, having been latterly confined to Araucarian fossils from the Mesozoic and Tertiary formations.

[4] Ludwig (1).

Without giving any list of the more recent fossil woods, five species may be mentioned from the *Potomac* (? *Wealden*) formation of America (Virginia). In all of these annual rings are recorded, and in three they are mentioned as distinct[1].

	Rings absent	Rings indistinct	Rings distinct
Cupressinoxylon pulchellum (*Knowlton*).			+
C. McGeei (*Knowlton*).			+
C. Wardi (*Knowlton*).			+
C. Columbianum (*Knowlton*).		+	
Araucarioxylon Virginianum (*Knowlton*).		+	

The genus Cupressinoxylon probably represents the wood of Sequoia.

In *Tertiary* fossil woods, so far as they have been examined microscopically, there are no very important points to notice with reference to annual rings. It is, however, interesting to note that Conwentz finds the annual rings in most of the Amber trees[2] of the Baltic coast normal in structure: in some he has detected double rings, and calls attention to the occurrence, in a few cases, of a "middle ring" between the normal autumn and spring wood; this he concludes may have been caused by a sudden leaf fall, the result either of atmospheric influence or destruction of the leaves by insects.

---

[1] Knowlton (1) and (2).

[2] The marine beds in which the Amber occurs are of lower Oligocene age; the trees lived probably in Eocene times. In his recent monograph [Conwentz (1)] Conwentz discusses at some length the formation of two rings of growth in one year.

CHAPTER VI.

Arctic Fossil Plants.

By far the most interesting lessons in questions of ancient climates have been taught by fossil plants found in the high northern latitudes of the Arctic regions. In 1840 Prof. Steenstrup[1] brought from Iceland a number of Tertiary plants which drew the attention of geologists to the evidence afforded by Arctic fossils of a much warmer climate having obtained in Polar regions. Seeing how important are the results which have been arrived at from a study of these northern plants in matters relating to geological climates, we may briefly sketch the main facts of circumpolar Fossil Botany and see on what data the conclusions as to climate have been based.

The late Prof. Heer has given a complete account of Arctic fossil vegetation[2].

In 1868 appeared the first volume of the *Flora Fossilis Arctica*; the seventh and last volume was completed in 1883, the last year of Heer's life.

Heer summed up in a few words the results furnished by the rich store of Greenland plants which formed the subject of his last volume. "[3] Cette riche collection nous a donné une connaissance exacte de la végétation qui, dans différentes périodes, a couvert le Gronland. J'ai réussi à reconstruire les forêts qui ont orné jadis ces régions du haut Nord, maintenant en partie couvertes de neige et de glaces; mais ce travail a en même temps fourni des matériaux du plus grand prix pour la détermination du climat pendant ces périodes, comme aussi pour l'histoire du développement du règne végétal et de sa répartition à la surface du

---

[1] Meddelelser om Grønland, Pt. v. 1883. *Resumé* by M. F. Johnstrup, p. 223.
[2] Heer (3).
[3] Meddelelser, loc. cit. p. 225.

globe. Même s'il reste encore beaucoup à éclaircir dans ce do-
maine, j'espère cependant avoir rendu quelques services à la
Science, et cela suffit pour me dédommager—je puis bien le dire—
du grand sacrifice que j'ai fait en vue de ce travail."

To give a full description of the vegetation of circumpolar
lands would be beyond the scope of the present work. Beginning
with the Palæozoic plants from Bear Island we will pass in review
the several floras from Arctic plant-bearing strata of different
ages[1].

Devonian and Lower Carboniferous.

Bear Island[2].   (Latitude 74° 30′ N.)

This small island, between Lapland and Spitzbergen, has
yielded a rich flora which Heer refers to his "Ursa Stage,"
occupying a position in the geologic scale between the Carboniferous limestone and Devonian rocks. Most of these plants were
collected by Nordenskiöld and Malmgren during the Swedish
expedition of 1868. Specimens were abundant, but species few;
18 species have been recorded, of which 15 occur in other
countries, three being peculiar to Bear Island.

Calamites radiatus, Lepidodendron Veltheimianum, Cyclostigma Kiltorkense and other familiar species are included in
this flora.

Spitzbergen.   (Latitude 78° 80′ N.)

From material collected by Nathorst and Wilander in 1870 on
the shores of Klaas Billen-Bai, Heer described five species, three
identical with Bear Island plants. These plant beds have been
referred to the Ursa Stage.

Melville Island.

From plants brought by M'Clintock to Dublin the existence of
Ursa beds in this island has been determined.

Greenland.   (*Disco Island.*)

In 1871 Fries and Nauckhoff collected a number of fossil
plants at Ujararsusuk; amongst them were several specimens of

---

[1] In addition to Heer, see also Feilden and de Rance (1), Rupert Jones (1),
and Ward (2), p. 826 et seq.
[2] Heer (3), vol. II.

Protopteris punctata which agreed closely with European forms. Heer[1] regarded this Tree-fern as proof of the Coal-measure age of these beds. After the publication of this determination Feistmantel wrote to Heer and shewed that the beds containing Protopteris in Bohemia had been erroneously recorded by Stenberg and others as Carboniferous, their real age being Cretaceous (Cenomanien). Heer, in correcting this mistake, remarks that there is no proof of the occurrence of Coal-measure fossils in Greenland[2].

Upper Carboniferous.

Asia.

On the banks of the Lena, six degrees from the Arctic circle, coal seams have been found with Calamites cannæformis[3], a European Coal-measure species.

Novaya Zemlya[4].

Specimens of the Coal-measure genus Cordaites were collected by Nordenskiöld in 1875, from Goose Cape. Heer considered the rocks in which the specimens were found as probably of Carboniferous age, but possibly they should be referred to the base of the Permian.

Spitzbergen[5]. (Latitude 77° 33′ N.)

In 1872 Nordenskiöld collected a number of plants in the Robert valley: these proved to be Upper Carboniferous species.

Heer mentions that many of the typical European forms are not represented in the Polar Carboniferous beds, for example, Calamites, Annularieæ, Asterophylliteæ, Sigillaria, Neuropterideæ and Pecopterideæ.

Jurassic.

Spitzbergen[6]. (Latitude 78° 22′ N.)

At Cape Boheman Nordenskiöld and Oberg collected plant remains of Jurassic age; Podozamites, Gingko and other European genera were found. These plant beds were formerly considered

---

[1] Heer (3), vol. ii.
[2] Heer (3), vol. ii. Meddelelser om Grønland, Pt. v. (1883), p. 74.
[3] Heer (3), vol. ii.   [4] Heer (3), vol. v.
[5] Heer (3), vol. iv.   [6] Heer (3), vol. iv.

to be of Tertiary age but were referred later by Heer to the Bathonien division of the Jurassic rocks.

Cycads were abundant and no species was found which would suggest a colder climate than obtained in Europe during that period. The Spitzbergen Jurassic plants afford, according to Heer, no evidence in favour of climatal zones.

Melville Island.

M'Clintock obtained a Conifer, Thuyites Parryanus, from a coal bed at Village Point, possibly of Jurassic age.

Cretaceous.

Spitzbergen.

In 1872 plants of Lower Cretaceous age were collected by Nordenskiöld from Cape Staratschin: 15 species were determined, 5 Ferns, 8 Conifers, 1 Monocotyledon. The plants are a mixture of Lower and Upper Chalk species.

Greenland[1].

The Cretaceous plants of Greenland have been referred to three stages:

(a) *Kome.* (b) *Atane.* (c) *Patoot.*

From the three horizons 335 species have been determined.

(a) *Kome.* [*Urgonien.*]

Several localities on the north side of the Noursoak Peninsula have afforded Lower Cretaceous plants, in all 88 species; 43 Ferns, 1 Rhizocarp, 1 Lycopodiaceous species, 3 Equisetaceæ, 10 Cycads, 21 Conifers, 5 Monocotyledons, 1 Dicotyledon. The plant-bearing strata attain a considerable thickness and rest immediately on gneissic rocks; they extend from 150 ft. above sea-level to a height of 2330 ft.

The Ferns form the chief part of the Kome flora, the Gleicheniaceæ being particularly prominent.

Next to the Ferns come the Conifers with Pinus and Sequoia as conspicuous genera.

The general character of this flora agrees with that of the Wealden vegetation.

---

[1] Heer (3), vols. VI. and VII. Meddelelser, Pt. v. (1883.)

(b) Atane. (Cenomanien.)

Plants of this age have been collected from 17 localities. Here the Dicotyledons come to the front; 90 species, in a total of 177, having been referred by Heer to this class. Ferns occur in abundance. Williamsonia cretacea is recorded from these beds; its first appearance in rocks later than the Jurassic.

(c) Patoot. (Senonien.)

The Patoot beds have yielded an abundant flora, with 116 species, which have been distributed as follows:—Fungi 1, Vascular Cryptogams 20, Gymnosperms 18, Monocotyledons 5, Dicocyledons 69. More than half the plants are Dicotyledons: Conifers are abundant and Cycads unrepresented.

Tertiary.

In describing the Tertiary floras of the Arctic regions it will be convenient to follow Heer as regards geological age, as it is from his writings that the great bulk of the facts relating to northern fossil floras have been drawn. The controversy as to the geological horizon to which the Arctic plant beds should be referred will be briefly alluded to later.

Spitzbergen.

Nordenskiöld and Blomstrand brought the first material, which was collected during the expeditions of 1858—61 and 64: additions were made in 1868 to the Tertiary plants.

A large number of species have been described by Heer[1] and regarded by him as Miocene. In describing the plant-bearing beds at Cape Lyell, Nordenskiöld[2] remarks—"The rocks on the coast for a distance of several hundred feet form a continuous herbarium, where every stroke of the hammer brings to light an image of the vegetation of a long-past age."

Iceland.

42 species of Miocene plants were determined by Heer from collections made by Steenstrup and Winkler[3].

---

[1] Heer (3), vol. II.   [2] Nordenskiöld (1).   [3] Heer (3), vol. I.

A more recent addition to our knowledge of Iceland fossil plants has been made by Windisch, who has given a general account of Tertiary plants from Iceland, and a description of new material collected by Dr Smith, of Berlin, in 1883[1].

Greenland[2]. (Latitude 69° 15′ to 72° 15′).

Several localities on the west coast have yielded plant remains of "Miocene" age; from 20 localities 282 species have been described by Heer; of these 9 are Cellular Cryptogams, 22 Vascular Cryptograms, and 251 Phanerogams. The three most abundant trees, which are widely spread in the Arctic Tertiaries, are Sequoia Langsdorfii, Taxodium distichum, and Populus arctica.

Johnstrup[3], in his *résumé* of the Greenland fossil floras, speaks of the enthusiasm displayed by Heer in the examination of the recently acquired collections; the sight of a splendid specimen of a Cycad leaf and the first fossil Palm from Greenland (Flabellaria grönlandica) caused him to manifest an almost childlike excitement.

Grinnell land. (81° 46′ N.).

During the expedition of H.M.S. *Alert* and *Discovery* Capt. Feilden and Dr Moss collected a number of Miocene plants near Cape Murchison which, for the most part, agreed with the species recorded from Greenland and Spitzbergen[4].

Lieut. Greely[5], in *Three Years of Arctic Service* notes the discovery of a fossil forest near Cape Baird in latitude 81° 30′ N.

Bathurst Island[6]. (75° 30′ N.).

In some coal at Graham Moore Bay, M'Clintock found one species, Pinus Bathursti, which is probably of the same age as the other Arctic Tertiary plants.

Banks Land[6]. (74° N.).

6 species were collected by M'Clintock and M'Clure to which Heer has assigned a Miocene age.

An abundance of fossil wood has been found in Banks land.

[1] Ward (2), p. 830.   [2] Heer (3), vol. vii. Meddelelser v.
[3] Meddelelser. *Resumé* (Pt. 5), p. 225.
[4] Feilden and de Rance (1).   [5] Greely (1), Vol. ii. app. xiv.
[6] Ward (2), p. 835. Heer (3), vol. i.

Prince Patrick Island[1]. (76° 15′ N.).

Large stems were discovered here in 1835 by Lieut. Mecham. Other explorers have collected fossil wood from the same locality.

Mackenzie River[1]. (65° N.).

In 1785 Sir Alex. Mackenzie discovered coal beds containing plant remains: these were noticed later by Franklin and Richardson, and the latter, in 1848, made a collection of plants which Heer examined and classed as Miocene. Additions have been made to the Mackenzie River flora by the Hudson Bay Company: the Mackenzie River district has yielded altogether 23 species of Miocene plants[2].

Quaternary.

Spitzbergen.

On the shores of Advent Bay a bed was discovered in 1868 containing a number of recent shells, also Mosses, with some species of Algæ and Equisetum variegatum.

We have seen that the Arctic regions have yielded an abundance of fossil plants of various geological ages. In enumerating the districts and countries from which plants have been collected no detailed lists have been given: the chief characteristics have in some cases been noted, and the points of difference between the Arctic and European floras will be mentioned in discussing Cretaceous and Tertiary climatal conditions within the Arctic circle.

So far as Palæozoic and Mesozoic plants are concerned there is no evidence, according to Heer, of any difference in the temperature of the Polar regions from that of Central Europe. The finds of pre-Cretaceous plants have been hardly sufficient for us to draw any very definite conclusions from them. All we can say is, that so far as the species which have been collected are concerned, no direct evidence is afforded in favour of the existence of climatal zones in Carboniferous or Jurassic times.

It has been already remarked that in the case of Coal-measure plants there is an absence, in such few collections as have been brought together, of several typical European species. Whether this must be regarded as the result of our imperfect knowledge,

---

[1] Ward (2), p. 835. Heer (3), vol. i.   [2] Ward (2), p. 835.

or as expressing a certain peculiarity in Arctic Carboniferous vegetation, we are not in a position to express a decided opinion.

Nordenskiöld[1], in a paper *On the former climates of the Polar regions*, notes the absence in Arctic Coal-measures of the large-leaved Ferns so abundant in the Upper Carboniferous rocks of more southern latitudes; this may, he thinks, possibly indicate a difference in the climate of Arctic regions, but more probably must be explained by the paucity of material collected.

To look a little more closely at the results arrived at from an examination of the Cretaceous and Tertiary floras of Greenland. At the time of the deposition of the Kome beds the climate of North Greenland was probably subtropical, and enjoyed a mean annual temperature of 21°—22°C. This estimate by Heer is based upon the occurrence of tropical and subtropical species, especially species of Gleicheniaceæ, Cycads, large-leaved Oleandras, delicate Adiantums and Aspleniums. The numerous Conifers, although not tropical, point to a subtropical or warm temperate climate. The more recent additions of Lower Cretaceous plants confirm the teaching of the original collections.

In his earlier account of the *Atane* beds, Heer[2] considered that the plants indicated a lower temperature in North Greenland than obtained in Bohemia and Moravia. The occurrence of only two Cycads and three Gleicheniæ suggested that a zonal difference in temperature had already been established in Europe. The plants collected since Heer's first descriptions were published have not supported his view, that at the time when the Atane beds were deposited there was a distinctly cooler climate in North Greenland than in Central Europe. Of the 177 species recently collected from these Atane beds, a large number are tropical and subtropical forms: Tree-ferns (Cyatheæ and Dicksoniæ), 6 Gleicheniæ, 8 Cycads, 7 Laurineæ, 2 Eucalyptus, &c., &c. Especially noteworthy are Dicksonia punctata and Cycas Steenstrupi; the latter is closely related to Cycas revoluta, which flourishes in the tropics, but is able to ripen its seeds in the gardens of Madeira. In New Zealand Dicksonia antarctica lives in latitude 43½°S. The Atane flora, Heer concludes, probably flourished in a climate with a mean annual temperature of 18°—19°C. : not cooler than the climate of Central Europe.

[1] Nordenskiöld (1).   [2] Heer (3), vol. vii.

In the *Patoot* flora, Cycads are not represented : a number of
temperate zone trees appear, such as Birch, Alder, Ash, Maple and
others; with these occur a few representatives of a warmer
climate, 3 Gleicheniaceæ, 2 Dammara, 2 Fig trees, 3 Sapotaceæ,
&c. The presence of these more southern plants indicates a higher
temperature for the Upper Cretaceous than for the succeeding
Tertiary period. This mixture of temperate and subtropical plants
in the Patoot (Senonien) beds forms a connecting link between
the tropical and subtropical Lower Cretaceous and the cooler
Lower Tertiary.

From his examination of the Miocene plants from Greenland
in 1868, Heer[1] estimated the mean annual temperature for North
Greenland (70°N.) at 9°C. at least ; the mean winter temperature
not below 0°C., and the summer temperature at $16\frac{1}{2}$°—$17\frac{1}{2}$°C.
The more recently acquired specimens led Heer[2] to raise the mean
annual temperature to 12°C. In comparing the Miocene and
recent plants, three groups may be noted. 1. In the first group
are included species whose living representatives are still found in
the Arctic zone : Pteris Oeningensis (represented by Pteris Aqui-
lina), Aspidium Escheri (represented by A. thelypteris), Populus
Richardsoni (represented by P. tremula), &c.

2. In group 2 are included species whose living representatives are outside the Arctic circle, but characterise places where
the mean annual temperature is about 8°—9°C. Onoclea sensi-
bilis, Osmunda Heerii, Gingko adiantoides, Thuya borealis, also
species of Quercus, Fagus, Betula, Acer, Liriodendron, Vitis, &c. are
examples of this group.

3. The members of the third group require a higher temperature. Flabellaria Grönlandica, F. Johnstrupi, belong to this group.
For these Heer considers we must allow a mean annual temperature of about 10°C., a winter temperature of 4.4°C., a summer
temperature of 16.3°C. This estimate is suggested by the conditions under which such species as Chamærops humilis, C.
Fortunei and Sabal Adamsonii, are able to live at the present
day.

The majority of the Miocene plants of Greenland correspond to
species living in the temperate zone, and which do not require a
higher temperature than 9°C. In Grinnell land (81°.44′N.) the

---

[1] Heer (3), vol. I.   [2] Heer (3), vol. VII.

occurrence of the Swamp Cypress and other foliage trees points to a temperature of about 8°C. : plants from the Eisfiord—Spitzbergen[1] —point to a mean temperature of about 9°C. To come further south we find a Miocene flora in Switzerland, in latitude 47°N., which is considered to have required a temperature of 20·5°C.; this gives for the Miocene period a decrease of ·37°C. in temperature for each degree of latitude, passing from central Europe to the Arctic regions. According to this estimate the temperature of North Greenland (70°N.) in Miocene times was about 12°C.

"*Are the fossil floras of the Arctic regions Eocene or Miocene?*" In an article in *Nature*, vol. xix. (1878), Mr Starkie Gardner[2] discusses this question and brings forward arguments for assigning the Arctic plant-bearing beds to the Eocene and not to the Miocene period, as Heer has done in his descriptions of Arctic floras.

Gardner points out that floras which are much alike and are met with in widely separated latitudes cannot have been contemporaneous; floras of quite distinct family may have been contemporaneous. In his anniversary address of 1862, Professor Huxley[3] quotes De la Beche and Edward Forbes in support of this same assertion, that "similarity of the organic contents of distant formations is *prima facie* evidence, not of their similarity, but of their difference in age." Gardner proceeds to bring forward physical evidence in favour of the Eocene age of the floras. During Eocene times land existed in the Polar regions, and one would therefore expect to find Eocene plants in their proper sequence. The temperature of the Eocene period in Europe was much higher than that of the Miocene period, and therefore more favourable to the existence of such floras in high northern latitudes. If, as Heer maintains, the Arctic plants are of Miocene age, there must have been at that time a climate uniform enough to have supported the same species of plants simultaneously from Italy and the United States to the 70th parallel of latitude. If the Arctic floras are of Eocene age, the decreasing temperature of Miocene times would drive the Eocene plants towards the south.

Gardner makes some general remarks on geological climates in

---

[1] There is an absence in Spitzbergen of such Southern forms as the evergreen Magnolia, Castanea, Prunus, Ilex, &c. which occur in Greenland.

[2] Gardner (1).　　　　　　　[3] Huxley (1), p. 21.

Arctic regions. He is of opinion that the tendency is to assume a higher temperature than is necessary for the growth of the floras, whose remains are found in Arctic plant beds in Northern regions. The great length of summer day is favourable to plant growth; the chemical action of the sun's rays compensating for feeble warmth. The remarkable capacity of Arctic and Alpine plants for living through extreme cold must not be lost sight of in speculating upon the necessary temperature for the Tertiary floras.

In the discussion following the reading of a paper by Captain Feilden and Mr De Rance on *The Geology of the Arctic Coasts*, Mr Belt[1] criticised the conclusions of Heer as to the Miocene age of the Arctic plants; he preferred to consider the supposed Miocene plant as Eocene, which lived during the Eocene period within the Arctic Circle, but in Miocene times migrated southward, when the climate of Arctic regions became too cold for them.

Mr Starkie Gardner, in support of the same view, drew attention to the fact that many Miocene plants were common to the Eocene, as had been shewn by species collected from American beds.

To these criticisms Heer replies in the seventh volume of the *Flora Fossilis Arctica*[2]. He holds that it is quite conceivable that Eocene beds may be unrepresented in the Arctic regions, but that in the great thickness of beds between the Lower Miocene and Upper Chalk in North Greenland there is no reason why Eocene fossils should not be discovered. He argues that the Northern flora, in its general character as well as in its specific character, agrees closely with European Miocene plants, and in following the flora from polar lands to South Europe there are no places where a commingling of Eocene and Miocene species is detected. Again, there is a great contrast between the Arctic floras and the European Eocene floras. In a paper on the leaf beds and gravels of Ardtun, &c., Starkie Gardner discusses at length this vexed question of age. He criticises severely what Robert Brown designated "the reckless way in which Heer makes species and genera" from fragmentary specimens[3].

Without attempting to follow Gardner's argument, we may

[1] Feilden and de Rance (1).
[2] Heer (3), vol. vii. p. 212.
[3] Gardner (2), p. 295 [Footnote].

note his conclusions. He considers the Arctic floras demonstrate a gradual passage from Cretaceous to Miocene floras. "The stratification of these beds renders it in the highest degree improbable that beds of Eocene age should be unrepresented, whilst the known temperature of the Eocene would have been more favourable to the growth of temperate floras in high latitudes than the diminished heat of the Miocene[1]."

Whether the Tertiary Arctic plants be of Eocene age, as Gardner and many geologists believe, or of Miocene age, as Heer persistently maintained, and allowing that many of Heer's determinations were based on too imperfect data, and the temperatures assigned to the Tertiary climate too high, we have sufficient evidence of very different climatal conditions from the present having obtained in Arctic regions. "When we remember," writes Geikie[2], "that this vegetation grew luxuriantly within 8° 15′ of the North Pole, in a region which is in darkness for half of the year, and is now almost continuously buried under snow and ice, we can realize the difficulty of the problem in the distribution of climate which these facts present to the geologist."

In one of the appendices to Nansen's recent book[3] on Greenland we find the following statement:—"The rocks in places give us indisputable testimony that its soil was once covered with luxuriant forests of palms and other tropical plants that we must now go to the latitude of Egypt to find." The evidence hardly justifies us in accepting this restoration of climatal conditions in Greenland in Tertiary times: indeed Nansen himself reminds us that such a high authority as Nathorst considers Heer's temperature estimates too high. Heer himself admits that the Tertiary climate of Greenland was cooler than that of more southern climates.

[1] Gardner (2), pp. 298—299.   [2] Geikie A. (1) p. 869.
[3] Nansen (1).

CHAPTER VII.

CARBONIFEROUS PERIOD.

In the references to the different views held on the subject of past climates which have been given in the historical sketch, we have seen how much attention has been paid to the Carboniferous period, and how general has been the opinion that the Coal period climate was mild, if not tropical, and very uniform; not only in the absence of any great variation in temperature during the year, but uniform also in the sense of geographical distribution, without any indications of climatic zones or botanical provinces. The suggestion too of an abnormal amount of carbonic acid gas in the atmosphere of the Carboniferous forests has had many supporters.

To consider, first, some of the arguments which have led to these conclusions, and then discuss somewhat fully the view recently advanced that there is reason to believe the climate was by no means tropical, and that evidence is not wanting in favour of the existence of Carboniferous or Permo-Carboniferous botanical provinces.

An abstract of the views held by Grand'Eury, Saporta and others will serve to shew the main arguments brought forward to support the opinions generally held as to the Coal-period climate.

Grand'Eury[1] does not overlook the difficulty of drawing conclusions as to climate from analogy, in the comparison of fossil and recent plants. The Marattiaceæ—so abundant in Carboniferous times—may, he suggests, have been capable of living under very different conditions than those most favourable to the present members of the family. The flora of the Coal-measures he con-

---

[1] Grand'Eury (1).

siders to be of a tropical nature. The abundance of Ferns suggests a resemblance to the warm and moist conditions which obtain in islands where Ferns reach their maximum development, and where Cryptogams are abundant. He deals at length with the inferences to be drawn from the nature of the plants, and mentions heat, moisture and light as the three principal factors which have most effect upon the organisation of plants.

Heat and Moisture.

Knowing that heat and moisture accelerate vegetation, one may infer, he maintains, that both were present in considerable abundance during the Upper Carboniferous period; this conclusion he considers warranted by the active and vigorous growth of which the coal seams afford striking evidence. The succulent nature of the Sigillarieæ, insisted upon by Lindley, is quoted by Grand'Eury as further proof of a high temperature and much moisture. This argument, advanced by Grand'Eury and others, can hardly be considered to have any great weight : in the first place, does the structure of the Sigillarieæ suggest a succulent character? And, moreover, if the Sigillarieæ were succulent plants, in the sense in which botanists are accustomed to use the term in speaking of recent plants, that would by no means be a proof of necessarily a moist climate. At the same time, the presence of much water in plant organs above the surface of the ground may indicate, as Sorauer[1] has shewn, a moist atmosphere. The dense foliage is considered to point in the same direction, and the large size of the vessels of some of the Carboniferous plants, previously noted by Corda, is included in the arguments in favour of tropical conditions.

This tropical climate was probably also a moist one, seeing that the marsh plants of warm countries have a considerable development of soft tissues ; for example, the large Cryptogams and arborescent Gramineæ of the tropics. No plants require a uniform and moist climate more than Ferns, and we know how abundantly this group of plants is represented in the Upper Carboniferous rocks. It is known that warmth and moisture together favour the elongation of plants, and, even in this particular alone, the Carboniferous vegetation denotes the influence of moisture and warmth

---

[1] Sorauer (1).

to a high degree. " Aerial roots " are mentioned by Grand'Eury as characteristic of a tropical and moist climate ; he refers to their great development in the Caulopterideæ and Calamitæ, an opinion, however, not shared by Prof. Williamson, as regards the Calamitæ at any rate.

The temperature was not, however, excessive, because the structure of the Coal plants does not permit of the supposition that they grew under very different conditions to those which at present obtain on the earth's surface. Unger has estimated the mean temperature at 20°—25°C., corresponding to that of islands in the South Seas.

Light.

The Sun's light, Unger observes, was probably obscured by vapours. This shutting off of light possibly accounts for the absence of Phanerogams. At the same time the light which governed the phenomena of nutrition of this luxuriant vegetation must have been very abundant, if not intense, considering the vigorous foliar development and active respiration and transpiration. After further remarks on the effect of light, Grand'Eury proceeds to argue that intense light is rather harmful than beneficial to the growth of plants. In a strong light they remain small and woody ; in shade they are less woody and more vigorous.

Experiments have shewn that the decomposition of carbonic acid gas is more considerable under the influence of sunlight lessened by a screen, than under the direct action of light.

In support of a uniform climate Grand'Eury refers to the absence of true rings of growth in Calamites; and agrees with Binney and others that the so-called rings described by some observers are due merely to differences in colour, and not to any variation in the actual size of the vessels. Such indications of annual rings as are found in Coniferous wood from the Coal-measures he regards as expressions of local disturbing influences on growth and not as proofs of seasonal changes.

Lesquereux[1] has expressed himself on the subject of the forma-

[1] Lesquereux (1).

tion of Coal differently to the majority of his contemporaries. He shews how peat bogs illustrate in all important points the process of the conversion of plant *débris* into coal. Coal, he says, must be considered the result of the alteration of half-aerial, half-watery vegetation, exactly corresponding to the formation of our peat bogs. In his work on peat, Lesquereux has compared in detail coal and peat bogs; they agree in their geographical distribution as well as in the character of their vegetation. We must note that in the comparison of the plants in the two formations a number of the Coal-measure plants are spoken of as belonging to other families than those to which they have been assigned since Lesquereux's account of peat was published. He defends his statement as to the similar distribution of peat and coal, as this opinion is not generally allowed by workers in Palæobotany.

Peat bogs, he shews, do not, as is generally asserted, belong especially to cold climates; they occur in the same latitudes as do most of the coal seams. The so-called "coal" of Brazil and Texas, which seems to afford an argument against the close analogy between coal and peat, he considers to be lignite and not true coal.

The fact that large tree stems occur in coal is a point of agreement with peat, and must not be considered an argument in favour of tropical conditions at the time of the Coal forests. The peats of Denmark and of the great Dismal Swamp have afforded large trees, equal in size to any of these in coal seams. Tree-ferns occur in the Coal-measures, and are by many regarded as proofs of tropical conditions. Lesquereux admits that Tree-ferns are characteristic of the tropics, but, he adds, they flourish in marshy grounds or on the borders of shallow lakes, and their position is such as to suggest that a moist atmosphere, rather than a tropical temperature, is the chief factor in determining their distribution.

Large Lepidodendra do not necessarily indicate a warm climate: in the peat bogs and forests of the northern hemisphere we find large Lycopodiaceæ represented in abundance.

On the whole the conclusion arrived at is that there is no evidence to warrant the assumption that atmospherical influences were very different from what they are now. "It is, then sufficient," he concludes, "to give us the reason of the differences in the type of vegetation between the Coal period and our own to admit that continents were less extended, and only low islands entirely covered with marshes[1]."

Saporta[2], in an article in the *Revue des deux mondes*, discusses the formation of coal according to Grand'Eury's theory, and deals with the question of Carboniferous climate.

The number of Upper Carboniferous Ferns, which were arborescent and rich in foliage, suggests a tropical climate, a sky frequently cloudy, and an atmosphere probably dense enough to obstruct the bright glare which nowadays both Ferns and Lycopods dread. The large extent of green surfaces, the abundance of parenchymatous and succulent tissues, and the insignificant size of the woody cylinders of the trees are all facts, Saporta notes, on which Grand'Eury's conclusions were based.

In *Le monde des plantes avant l'apparition de l'homme*[3] Saporta repeats his belief in a uniform Carboniferous climate. The origin of peat is discussed as calculated to throw light upon the formation of coal. Peat mosses require a uniform and not a high temperature, moisture, a flat country, and an impermeable soil. If, adds Saporta, we picture to ourselves the physiography of Upper Carboniferous times, and add to this a moist heat, a dense atmosphere charged with aqueous vapour, and a climate with violent and frequent rains, the Coal flora loses much of its singularity, and the method of formation of coal is naturally explained.

Sir William Dawson[4] speaks of the Carboniferous climate as moist and equable, but not tropical. He too adds a caution against drawing conclusions as to temperature from extinct genera. In describing Sternbergia (Artisia), he adds—"As Renault well remarks with reference to Cordaites, the existence of this chambered form of pith implies rapid elongation of the stem, so that Cordaites and Conifers of the Coal formation were probably quickly growing trees."

Credner[5] describes the climate of the Coal age as moist, warm and without frost. That this climate obtained in all latitudes is shewn by the occurrence in Equatorial districts, and in Arctic regions, of the same Carboniferous plants. He considers we may legitimately assume a larger percentage of carbonic acid gas to

---

[1] Lesquereux (1), p. 845.   [2] de Saporta (6).   [3] de Saporta (1) and (2).
[4] Dawson (2), p. 137.   [5] Credner (1).

have been present in the atmosphere of that period than in the atmosphere of to-day.

Mr John Ball[1], after calculating the amount of Coal in the earth, and the amount of carbon which must nearly all have been extracted from the atmosphere, comes to the conclusion that during Palæozoic times, before the deposition of the Coal-measures, the atmosphere contained twenty times as much carbonic acid gas and much less oxygen than now.

In a recent address, delivered by Mr T. V. Holmes[2] before the Geologists' Association, we read—"As regards the Coal-measures few, if any geologists now believe that there was any unusual amount of carbonic acid gas in the atmosphere of the world generally during their deposition." In speaking of the similarity of widely separated coal plants Mr Holmes suggests that the significance of this does not appear so great if we remember that these are simply the representatives of floras which grew in swamps.

Prof. Zeiller[3], after enumerating certain Carboniferous plants from the coal basin of Tete in Africa, shews how clearly they correspond with the Upper Carboniferous plants of Europe and other regions: he regards the occurrence of the same specific types in the Coal-measures of Arctic, Temperate and Equatorial regions as suggesting an absolutely uniform climate.

Prof. Neumayr[4], in a lecture delivered at Vienna in 1889, makes an attack upon the current ideas regarding climates of past ages. Referring to the once widely accepted belief in a uniform climate in the earlier geological periods, due to the greater internal heat of the earth, he remarks—"The falsity of these assumptions is now pretty generally recognized, and the number of their adherents diminishes daily." Neumayr then proceeds to analyse the methods employed in climate discussions; he instances fossil plants of European Tertiary strata as affording a case in which evidence of a warm climate is undeniable, but goes on to shew how a universal extension of similar reasoning "leads to deceptive results, and the whole method must be employed with the greatest caution." As Neumayr's remarks have a most important bearing on the present subject no apology is needed in quoting his words at some length. Having briefly sketched the

[1] Ball (1), cf. also Hunt, T. Sterry (1).
[2] Holmes (1).   [3] Zeiller (2).   [4] Neumayr (3).

leading characters of Upper Carboniferous vegetation, he proceeds,
—" The geographical extent of this typical flora was extraordinarily
great; we trace it from the shores of the Atlantic through the
northern half of the old World to China, and it is also greatly
developed in the Eastern half of the United States. There, and
in China, are the greatest developments of beds of coal. Besides
these we find similar deposits with nearly the same vegetation in
the far north, in the American polar archipelago, in Spitzbergen,
and Nova Zembla. It is these facts that have led to the conclusion, already mentioned, that in the Carboniferous period a
uniform climate prevailed from the equator to the pole, together
with a dense atmosphere rich in carbon-dioxide, and impenetrable
to the solar rays. And yet a simple examination of the facts
assures us that all these suppositions are groundless. In so far as
regards the character of the flora, we really know nothing of the
temperature requisite to the Calamites, Lepidodendra, Sigillarieæ
and other extinct types." Similar views have been expressed by
Prof. Judd[1]. "Just as little reason is there for inferring that
Sigillarids, Lepidodendrids, and Calamites could only have lived
in tropical jungles, as there is for the once popular notion that
they flourished in an atmosphere supplied with a very exceptional
proportion of carbonic acid! Conifers grow now in very severe
climates, and only the Tree-ferns really indicate warm climatic
conditions. At the present day their chief development is in the
tropics, and they require, not indeed great heat, but the absence
of frost. We do not, however, know that this was equally the
case in former ages; in the Carboniferous period, the highest
division of the vegetable kingdom, now so dominant, the flowering
plants, were either non-existent, or were sparsely represented only
by a few early forms, and it is by no means improbable that these
types in their gradual extension have exterminated the Tree-
ferns in the colder regions to which they formerly extended, and
that these latter have lost the power which they once possessed
of withstanding frost."

The argument for a warm climate drawn from the great thickness and extent of coal seams is assailed, and it is shewn that a
luxuriant vegetation is by no means a proof of tropical conditions.

[1] Judd (1), p. 72.

Peat is the present-day example of an accumulation of vegetable matter corresponding in all probability to the conditions under which the *débris* of Carboniferous forest gave rise to coal. " It is not," says Neumayr, "in the towering primæval forest of India and Brazil, nor the Mangrove swamps of tropical coasts, but in the moors of the subarctic zone, that plant remains are now being stored up in a form that, in the course of geological ages, may become converted into beds of coal."

Had the climate of our latitudes been tropical the coal plants would have decayed too rapidly, and the conditions favourable to such preservation as would lead to the formation of Carboniferous peat would only be obtained in a temperate climate. The fact that typical Upper Carboniferous plants have been found as far north as latitude 76° N., compels us to assume a somewhat higher temperature in polar regions than at present obtains in high northern latitudes, and we may further conclude that in the temperate zone a cool climate obtained; in the summer moderate heat, in the winter moderate cold, but little frost. It is pointed out that the most extensive coal deposits are all in the temperate zone and " chiefly concentrated in its middle and northern regions." Although the typical Upper Carboniferous plants of Europe are absent in the tropics, other plants have been found in South Africa, India and Australia of quite another type constituting what has been called the *Glossopteris flora*. With the strata in which the members of this Glossopteris flora first appear are associated what many geologists regard as proof of ice-action. " From the facts we have recounted," says Neumayr, " bearing on the climate of the Coal-measure period, it is abundantly manifest that everything runs counter to the assumption of a uniform and warm terrestrial climate from the equator to the poles. Geographically we have sharply contrasted floras, and we have, moreover, widely distributed deposits, in the formation of which great masses of ice must have played a part, and thus the old views are utterly overthrown. We may say with much probability that the differences of the floral regions must be ascribed to differences of climate, and that, locally, the temperature was so low as to allow of the formation of great masses of ice; but anything beyond this is quite uncertain, and no one of the assumptions that have been made to explain the conditions of that epoch has any claim

to validity." Stur[1] has also connected the Glossopteris flora with
the ice period at the end of Carboniferous times.

The Coal-measure plants have without doubt a wide geographical distribution, but there are various facts to take into
account which must be considered in any attempt to use as
thermometers these plants from the Carboniferous forests. It
must not be forgotten that in Palæozoic times, so far as we know,
none of the highest plant forms were in existence, and the elements
of the Carboniferous floras had not these powerful opponents to
compete with in the struggle for existence. Again, the nature of
the Coal-measure plants was such as to insure, by means of spores,
a wide distribution. A more careful comparison of Carboniferous
plants from different countries and latitudes shews that the uniformity of the vegetation of that period does not appear to have
been quite so striking as is frequently asserted.

The Carboniferous flora of North America contains a large
number of species which are not represented in the European
flora; indeed Rothpletz[2] considers about half of the North American species are absent in Europe.

In China, again, although some of our European plants have
been recorded, one of our commonest genera, Sigillaria, is extremely
rare. The Gymnosperms were largely represented in the Carboniferous vegetation of China, and held a much more prominent
position than in Europe. A list of the Carboniferous plants of
China is given by Schenk[3] in the third volume of Richthofen's
China.

The greatest deviation from the normal type is found when we
examine the lists of plants from Carboniferous or Permo-Carboniferous beds in South Africa, India and Australia. There we have
the Glossopteris flora. This brings us to the questions—1. Of
what age are the Glossopteris-bearing beds? 2. Is there any
evidence in these countries in support of the assertion that certain
strata associated with the Glossopteris beds owe their origin to
wide-spread ice-action?

These questions must be considered in some detail, as upon
the answers to them depends the force of Neumayr's arguments in
favour of the existence of distinct floras in the Carboniferous
period.

It will be more convenient to first describe separately the rocks of India, South Africa, and Australia, so far as they bear upon these questions, and then discuss their correlation.

India.

The beds in India which concern us in the present discussion on the Glossopteris flora and its relation to the Carboniferous or, better perhaps, the Permo-Carboniferous vegetation of Europe, are included in the *Gondwana system*, the most important and widely-spread formation in the Peninsula. The rocks included in this system are considered to represent a much longer period of geological time than any of the typical European systems: they are probably of freshwater or fluviatile origin. In the *Manual of the Geology of India*, by Medlicott and Blanford[1], a twofold division is assigned to this system, Upper and Lower: Feistmantel has instituted a three-fold division : we are only concerned with the Lower Gondwana rocks, as defined by the authors of the *Manual*. The strata of the Lower Gondwanas attain a thickness of about 13,000 feet, and include the following series in descending order :

$$\left\{\begin{array}{l}\text{Panchet.}\\\text{Damuda.}\\\text{Talchir and Karharbari.}\end{array}\right.$$

These series we will describe very briefly.

Talchir Series[2].

A series of clays and sandstones. In some beds blocks of metamorphic rocks abound, also pebbles with striated surfaces. The clay strata, containing the boulders, rest in some places on a polished floor of older rocks. A few fossils are recorded from the clays, viz. :—

> Schizoneura sp.
> Gangamopteris cyclopteroides, *Fstm.*
> G. angustifolia, *McCoy.*
> Glossopteris sp.
> Noeggerathiopsis Hislopi. (*Bunb.*)

In the *Karharbari series* seams of coal occur, containing plant

---

[1] Medlicott and Blanford (2).   [2] Blanford, W. T. (4), Waagen (2).

remains : these beds stand in closest relation to the Talchir beds, and can hardly be regarded as a separate division. The following fossils are recorded from these beds :—

> Schizoneura, cf. Meriani, *Schimp.*
> Vertebraria Indica, *Royle.*
> Neuropteris valida, *Fstm.*
> Gangamopteris, 4 species.
> Glossopteris, 2 species.
> &c., &c.

Damuda.

This series is made up of sandstones, clays, and seams of coal. Ferns make up a large proportion of the flora, and more than half of them are forms with entire and reticulately-veined leaves. Ferns and Equisetaceæ play the most prominent part in the vegetation, whilst Gymnosperms occupy a subordinate position. The following genera have been mentioned :—

> Glossopteris, 16 species.
> Gangamopteris, 4 species.
> Schizoneura Gondwanensis, *Fstm.*
> Phyllotheca, 2 species.
> Alethopteris Whitbyensis, *Göpp.*
> &c., &c.

Panchet.

This series consists chiefly of sandstones : fossils are rare.

The Salt-Range boulder beds[1] (*North-West India*).

It would serve no useful purpose in the present discussion to consider at length the important facts brought to light by the geological survey of the Salt-Range district of Northern India, or to enter into the details of the controversy as to the age and correlation of certain boulder beds found in the eastern and western parts of the range. We may note, however, that in the eastern boulder beds, which are included by Wynne[2]—whose memoir on the district should be referred to for further details—in the

---

[1] For particulars as to the Salt-Range district, and the controversy as to the age of the boulder beds, the following references may be consulted, Oldham (3), Medlicott (1), Warth (1), Waagen (1), (4), (5), Blanford, W. T. (5), Feistmantel (5).

[2] Wynne (1), (2).

Cretaceous system, nodules have been discovered containing the
following Carboniferous fossils :—

> Bucania, cf. Kattaensis, *Waagen.*
> *Conularia lævigata, *Morr.*
> *   „   tenuistrata, *McCoy.*
>   „   cf. irregularis, *Kon.*
> *Aviculopecten, cf. limæformis, *Morr.*
> *Spirifer vespertilio, *Sow.*
>   &c., &c.

* These are identical with Australian Carboniferous species.

Some geologists have contended that these nodules occur *in
situ*, others consider them derived. Although we cannot here discuss this subject at length, it should not be entirely omitted in
view of its important bearing on the question of the age of the
Talchir beds, to which the Salt-Range boulder deposits bear a very
close resemblance. If the Salt-Range beds and the Talchir beds
are of the same age, we derive important evidence from the former
as to the age of the latter, and further extend the geographical
distribution of the southern hemisphere glacial deposits[1]. In
any case, as Medlicott[2] points out, if the Carboniferous age of
the eastern boulder beds is not established, " those of the western
and trans-Indus sections, which are undoubtedly palæozoic, will
still hold the position assigned to them by Dr Waagen, as pre-
sumably representing the Talchirs."

Before considering the question of geological age, it will be
better to give a brief summary of the rocks of Australia and South
Africa, to which references will be made in discussing the general
question of the Glossopteris flora.

Africa.

The African deposits have not been studied to the same extent
as the Lower Gondwana beds of India. The great *Karoo formation*
extends over a large area in the northern part of Cape Colony, the
Orange Free State, and the Transvaal : it is made up of a succes-

---

[1] Blanford, W. T. (5), pp. 257—258. Discoveries are here alluded to by Oldham
and Griesbach which still further extend the geographical limits of the representatives of the Palæozoic Talchirs.

[2] Medlicott (1), p. 133.

sion of sandstones and clays, with some coal seams and eruptive rocks. The Karoo beds rest on sandstones (Table Mountain sandstone), containing in some places typical Carboniferous plants, such as Lepidodendron and Calamites, which, according to Professor Zeiller, correspond to the Middle Coal-measures of Europe. W. T. Blanford[1] and Dunn[2] describe the Karoo beds as conformable to the underlying plant-bearing Palæozoic strata. Moulle[3] and others consider the two series unconformable.

The classification of the South African rocks is by no means definitely settled : several systems have been proposed by various writers, Wyley, Dunn, Stowe, Rupert Jones, Green[4] and others. Waagen[5] in his *Carboneiszeit* adopts the classification proposed by Wyley in 1867.

The following classification is that adopted by Professor Green[6].

Ecca beds.

Dwyka Conglomerate.

(Unconformity).

Quartzites of the Wittebergen, Zuurbergen, and Zwartebergen.

Bokkeveldt beds.

Table Mountain sandstone.

The *Ecca (Dwyka) conglomerate,* which generally overlies the older rocks, consists of blocks of Granite, Gneiss and other igneous rocks embedded in clay. We shall revert later to the question of the origin of this conglomerate. The conglomerate rests, as a rule, on the Table Mountain sandstone, which is often marked on its surfaces with grooves and scratches.

The *Ecca beds* include Coal Seams, in which Glossopteris has been found.

The upper part of the Karoo system need not be described, as it does not affect the question at issue.

Australia.

Here again we find much confusion in the classification and geological age of the strata with which we are at present concerned[7].

---

[1] Blanford, W. T. (4).　　　[2] Feistmantel (5), pp. 43, 45.　　　[3] Moulle (1).
[4] Green (1): see also Feistmantel (5).　　　　　　　　　　　　　　[5] Waagen (2).
[6] Green (1), p. 240.　　　　　　　　　[7] Feistmantel (6).

Eastern Australia.

The Carboniferous beds usually rest unconformably upon the older rocks (Granite, &c.). In the interior occurs a succession of sandstones which have yielded Lepidodendron and Cyclostigma; these beds may be of Devonian age.

Clarke[1] classifies the Carboniferous beds as follows:

> Wianamatta beds.
> Hawkesbury beds.
> Newcastle beds.
> Muree beds { Upper marine beds. / Older Coal seams. / Lower marine beds.

Muree beds. These beds are especially important as they afford, according to many geologists, distinct proofs of ice action. The series consists of coarse conglomerates and boulder deposits, including subordinate shales, sandstones and coal seams. These beds are best exposed at Stony Creek and near Greta in cuttings of the Great Northern railway: they have yielded the following plants :

> Phyllotheca sp.
> Glossopteris, 5 species.
> Noeggerathiopsis prisca, *Fstm.*
> Annularia Australis, *Fstm.*

The marine beds, with which these plant beds are associated, have afforded a number of shells which have been described by de Koninck[2] as characteristic Carboniferous fossils.

In other parts of the country, near Strond, Arowa, Port Stephens and Smith's Creek another flora occurs, older than that from Stony Creek, including Calamites radiatus, Sphenophyllum sp., Rhacopteris (4 species), Archæopteris Wilkinsoni, Cyclostigma Australe, Lepidodendron Veltheimianum, &c. These beds are probably of Carboniferous limestone age.

Newcastle beds. These consist mainly of sandstones, shales and coal seams, and are closely connected with the underlying series. A number of plants have been recorded, of which the following may be mentioned:

----

[1] Clarke (1), (2).   [2] de Koninck (1).

Phyllotheca Australis, *McCoy.*
Vertebraria Australis, *McCoy.*
Sphenopteris, 7 species.
Glossopteris, 7 species.
Gangamopteris, 2 species.
&c., &c.

Hawkesbury beds.

Here again are traces of ice action. The series is made up of sandstones, marls and conglomerates. A few plants have been found, among which were:—
　　　　Thinnfeldia odontopteroides, *Fstm.*
　　　　Sphenopteris sp.
　　　　Odontopteris sp.

Wianamatta beds. From shales and sandstones of this age the following fossil plants have been recorded:
　　　　Thinnfeldia odontopteroides, *Fstm.*
　　　　Odontopteris microphylla, *McCoy.*
　　　　Pecopteris tenuifolia, *McCoy.*
　　　　Macrotæniopteris Wianamatta, *Fstm.*

Queensland.

In Queensland Coal-measures are known belonging to two periods. The older contain marine fossils of a Carboniferous type, with Glossopteris, Schizopteris and Pecopteris. Traces of ice action have been noted.

The more recent Coal-measures have yielded the following plants:
　　　　Sphenopteris elongata, *Carr.*
　　　　Thinnfeldia odontopteroides, (*Morr.*)
　　　　Cyclopteris cuneata, *Carr.*
　　　　Tæniopteris Daintreei, *McCoy.*
　　　　　　&c., &c.

Victoria.

At the base of the Victorian series occur sandstones containing Sphenopteris iguanensis and Archæopteris Howitti, &c.; these are probably of Devonian age.

Above these beds come the Avon river sandstones in which Lepidodendron Australe has been found: these have been classed as Carboniferous. Next comes the Bacchus-Marsh sandstone with boulder deposits: Gangamopteris (3 species) has been found in this subdivision.

Above this sandstone come the Bellarine beds containing a Mesozoic flora.

Tasmania[1].

The beds of Tasmania may be classed as

> Upper Palæozoic (Lower Coal-measures).
>
> Upper Coal-measures (Mesozoic).

The position of these deposits will be shewn in the general table of correlation[2].

In dealing with the question of correlation of the Indian, Australian and South African beds we have to examine the arguments contained in a mass of controversial literature. The conflicting evidence of plants and animals has given rise to endless discussion, and all that will be attempted here is to point out some of the chief points of difference in the several opinions, and to give a summary of the results arrived at.

In correlating the strata of these different countries in the Southern Hemisphere the boulder beds already alluded to are important guides. H. F. Blanford[3], in a paper on the *Plant-bearing series of India*, goes so far as to suggest that in certain cases the physical evidence of glaciation becomes at least of equal value with palæontological evidence, perhaps even of greater value when it is a question of correlating formations in very distant regions. How far these boulder beds should be referred to ice action will be considered later.

After an exhaustive survey of the opinions of various writers on South African and Indian stratigraphy Feistmantel summarises the result as follows:

1. In South Africa occur Devonian and Carboniferous rocks.

2. The Carboniferous beds contain plant remains corresponding to European Coal-measure plants.

3. Above these Carboniferous beds are strata of the Karoo formation, containing fossils similar to those in the Damudas, Panchet and Upper Gondwanas of India.

---

[1] Johnston, R. M. (1), Feistmantel (7).   [2] See p. 122.   [3] Blanford, H. F. (1).

4. Between 2 and 3 come the Ecca beds, consisting of shales and conglomerates.

5. The Ecca beds are above the beds with Coal-measure plants, and are therefore younger.

6. The Ecca conglomerate (Dwyka conglomerate) is regarded as the result of glacial action. The Ecca beds are probably of Permian age.

It is only natural to assume that this great difference in the climatic conditions which gave rise to the boulder accumulations, has also had an effect upon the flora of the Carboniferous beds.

7. This Ecca (Dwyka) conglomerate must probably be correlated with the Talchir beds of India.

Considerable discussion has taken place as to the age of the members of the Gondwana system, especially the Damuda and Talchir series which most nearly concern us in the consideration of the Permo-Carboniferous flora. Feistmantel and Blanford have supplied most of this controversial literature. In his later papers, Feistmantel changed his opinion and, as shewn in the table quoted from him[1], assigned the Damudas and Talchirs to an earlier age than in his previous writings[2].

Without entering into tedious details of Australian stratigraphy, we may refer to Feistmantel's summary:

1. In Australia, in the Lower Coal-measures, occurs a marine Carboniferous fauna, and a flora (Glossopteris, Phyllotheca, &c.) which is abundantly represented in overlying beds.

2. The Upper marine beds and Newcastle beds in N. S. Wales, the Bacchus-Marsh beds in Victoria, the Talchir conglomerate in India, and the Ecca beds in Africa are analogous deposits, and represent the Permian period.

3. All these deposits contain boulders indicative of glacial action and point to a low temperature in the several districts.

4. The Indian Coal beds, the Damuda and Panchet series, the Lower Karoo beds in South Africa, and the Hawkesbury-Wianamatta beds in N. S. Wales are of Triassic age.

[1] See p. 122.   [2] Blanford, W. T. (1), (2).   Feistmantel (1), (2), (3), (4), (5).

Boulder beds.

The boulder beds which occur in India, South Africa, Afghanistan, and Australia have been regarded by some as of great weight in correlating the widely separated strata, and as evidence of wide-spread glacial conditions. It has also been suggested that the lower temperature, of which the boulder deposits afford proof, was no doubt the real cause of the apparently sudden change in the facies of the Southern Hemisphere vegetation in Palæozoic times. In reference to this suggestion as to the wide-spread glacial conditions, and the reaction of the low temperature upon the flora, we ought to consider the question as to the origin of these so-called " glacial deposits."

Talchir conglomerate.

These boulder beds were discovered by W. T. and H. F. Blanford, in Orissa, in 1856. The occurrence of boulders in fine silt was thought to be suggestive of ice action. A sketch is given in a paper by W. T. and H. F. Blanford, and Wm. Theobald[1] of the boulder beds. It is pointed out that probably ground-ice was the cause of the deposit. The writers note the possibility of a similar association of boulders and fine silt in some mountain lakes, such as Geneva, but add that such could only be of very local occurrence.

In 1872 Mr Fedden[2] discovered scratched boulders, and also striated and polished rock surfaces. The limestone on which the boulder beds rested was exposed for a distance of 330 yards, and shewed a scratched and polished surface, indicating the action of ice. Fedden concludes, "The evidence for the glacial origin of the deposit is as conclusive as that for the ice age formation in Europe."

The scarcity of fossils in the Talchir beds is considered by H. F. Blanford[3] another fact in favour of ice action.

Medlicott and Blanford[4], in the *Manual*, mention the fact that the Talchir boulders are very irregularly distributed, and in many cases have been transported from a distance.

---

[1] Blanford, W. T. and H. F., and Theobald (6), also Blanford, W. T. (3).
[2] Fedden (1). [3] Blanford, H. F. (1). [4] Blanford, H. F. and Medlicott (2).

Griesbach[1] describes the boulders as varying in size from peas to 30—40 feet in diameter, shewing evidence of having been deposited from above in a fine and soft matrix. He gives a coloured plate, shewing the Talchir conglomerate near Kandia. Scratched boulders from the Salt-Range are figured by Warth[2] and Waagen[3].

In the Kashmir area boulder beds, probably corresponding to the Talchirs, are recorded and considered by McMahon and Lydekker to be of glacial origin[4].

Ecca beds.

A glacial origin for these beds was first suggested by Sutherland[5].

Griesbach[6] gives a figure shewing boulders in clay : he notices that many are still angular, and have not therefore travelled far. He mentions too an abundance of grooves and scratches, similar to such as are seen on the rocks of the European Alps.

In a paper by Dr A. Schenck[7], on the geological development of South Africa, the Dwyka conglomerate is described as having a wide geographical extension, and containing an abundance of angular and rounded fragments of Granite, Gneiss, Schists, and other rocks. He mentions the view, held by Bain and Wyley, that the conglomerate was formed by volcanic agency; but notes that its glacial origin, as suggested by Sutherland and Dunn, has constantly gained in probability, and especially receives support from the close correspondence between the Dwyka, Talchir, and Bacchus-Marsh boulder beds. The same writer, in another place[8], repeats his opinion as to the glacial origin of the Dwyka conglomerates, and notes the occurrence of smoothed and scratched blocks, suggesting a strong likeness to the North German glacial deposits.

Australian boulder beds.

In 1866, Dr Selwyn and Sir R. Daintree found the Bacchus-Marsh boulder beds: they regarded the conglomerate as probably

[1] Griesbach (2).   [2] Warth (2), Pl. i. p. 34.
[3] Waagen (2), Pl. opp. p. 130.   [4] Lydekker (1).   [5] Sutherland (1).
[6] Griesbach (1).   [7] Schenck, A. (1).   [8] Schenck, A. (2).

due to marine glacial transport. In 1878 Mr R. L. Jack referred the boulders to ice action. Mr Wilkinson also mentions the glacial origin of the Australian boulder bed. In 1885 Mr R. Oldham discovered striated pebbles and boulders in the beds of Stony Creek, scattered through a matrix of fine sand or shale[1].

Afghanistan.

It may be noted that Griesbach has described conglomerates in the provinces of Herat and Khorassan, which he considers homotaxial with the Talchir boulder beds of India[2].

South America.

Dr Waagen[3] quotes a letter from Mr Derby, of the Rio geological museum, in which the occurrence of Lepidodendron, Cordaites and Psaronius is noted in certain beds in the Province of Parana. Mr Derby speaks of deposits which Waagen considers should be classed with the other boulder beds of the Southern Hemisphere. No scratched surfaces are recorded. The same observer admits that his usual custom is to make geological tours by rail; on one of his railway traverses he noticed rounded blocks, embedded in and protruding from fine clay slate. We cannot, however, attach any great importance to evidence based on passing glances from the windows of a railway carriage.

The two following tables, from Feistmantel and Waagen, are reproduced in order to facilitate a comparison of the Southern hemisphere rocks from the point of view of correlation. The boulder beds in the Salt-Range, and their important bearing upon the age of the Talchirs, have already been referred to.

[1] David (1). See also Dunn (1), and Oldham (1).
Blanford, W. T. (7). The boulder beds of India, Africa and Australia are described (p. 96) as "the best marked proofs, yet recorded, of glacial action in ancient rocks."

[2] Griesbach (3). Feistmantel (5), p. 93.   [3] Waagen (3).

GLOSSOPTERIS FLORA: CORRELATION OF STRATA. [Adapted from Feistmantel (7), pp. 616, 617.]

TASMANIA	EASTERN AUSTRALIA			S. AFRICA	INDIA	
	VICTORIA	N. S. WALES	QUEENSLAND			
Palæozoic Coal-measures (a) Upper Marine (*Gangamopteris angusti-folia*).	Bacchus-Marsh Sandstone, with *Gangamopteris*.	Newcastle beds, with *Glossopteris, Gangamopteris*, &c.	Upper Group, mostly fresh-water, with associated marine beds: *Glossopteris* abundant.	Ecca beds, *Glossopteris*, &c.	Karharbari Coal-measures and Talchir beds, with *Glossopteris*, &c.	LOWER GONDWANA = PERMO-CARBONIFEROUS.
(b) Tasmanite bed.						
(c) Plant beds, with *Glossopteris, Gangamopteris*, &c.	Bacchus-Marsh Conglomerate (glacial).	Upper Marine (glacial). Lower Coal-measures. Lower Marine.	Marine beds, with *Glossopteris* (glacial).	Dwyka conglomerate (glacial).	Talchir conglomerate (glacial).	
(d) Coal-measures.				Carbonaceous beds.		
(e) Lower Marine.	Avon Sandstone, with *Lepidodendron Australe*.	Strond and Port Stephens beds, with *Calamites radiatus, Lepidodendron Veltheimianum*, &c.	Drummond Range beds, with *Calamites radiatus*, and *Lepidodendron Veltheimianum*.	Table mountain Sandstone, &c.	Vindhyan formation.	LOWER CARBONIFEROUS.
Soft shales of Fingal (Freshwater), with *Anodonta Gouldi*.	Iguana Creek beds.	Goonoo-Goonoo beds, with *Lepidodendron nothum*.	Mt Wyatt beds, with *Lepidodendron nothum*.	Lower Cape formation, with Devonian fossils.		DEVONIAN.

GLOSSOPTERIS FLORA. [Waagen (2), p. 111.]

	EASTERN AUSTRALIA	S. AFRICA	INDIA
LOWEST TRIAS	Wianamatta beds. *Unconformity.*	Beaufort beds.	Panchet beds.
PERMIAN	Hawkesbury beds (glacial).	Koonap beds.	Damuda beds.
UPPER CARBONIFEROUS	Newcastle Beds. Stony Creek beds. Bacchus-Marsh beds (glacial).	*Unconformity.* Ecca beds (glacial).	Karharbari beds. Talchir beds (glacial).
LOWER CARBONIFEROUS	Strond and Port Stephens beds, &c. Lepidodendron beds.	Lepidodendron beds.	*Resting unconformably on crystalline rocks.*
DEVONIAN	Marine Devonian.	Marine Devonian.	

We have seen that facts are by no means wanting in favour
of the action of ice in the formation of the various "boulder
beds:" the evidence seems to point to glacial conditions. In
discussions as to the age of the Australian Glossopteris beds, it
has been pointed out that most of the expressions of opinion in
favour of the Permo-Carboniferous[1] age of these deposits are
from geologists who have had opportunities of examining the
rocks *in situ.* The majority of those who prefer to consider the
Glossopteris flora of a later age have only seen collections of plants
which have been brought to England[2]. The same fact might
be pointed out in discussions on the boulder beds: those who
have worked at these beds in India, Africa and Australia find
themselves driven to accept ice action as the only satisfactory
explanation of the origin of the widespread boulder deposits.
Many, on the other hand, who have seen a few specimens of
scratched and polished rocks brought from India are very sceptical
as to the value of the evidence in favour of ice work.

It is true, that, so far as one can judge from such figures as
those given by Warth[3] and Waagen[4], the appearance of the rock
fragments is not quite such as we are familiar with in the *moraine
profonde* of the European ice-sheet. Granting, however, this
difficulty, there still remains to be advanced a reasonable and
satisfactory explanation of the origin of these widespread boulder
beds by those who prefer to look to other agents than ice.

It is well known that the scratching and polishing of rocks
may be brought about by other forces than ice[5]; but none of the
explanations of such phenomena on a small scale can well be
applied to the boulder beds of the Southern hemisphere.

Taking into account the widespread distribution of the boulder
deposits, and the fact that they occur in several regions of the
Southern Hemisphere in association with Glossopteris, many
geologists have expressed the opinion that in these now separated Southern lands we have the remnants of a Southern Palæozoic continent.

Suess, in his recent *Antlitz der Erde,* supports the view in

[1] Blanford, W. T. (7).   [2] Blanford, W. T. (1), p. 83.
[3] Warth (2).   [4] Waagen (2).
[5] David (1). *Discussion*, p. 195. Blanford, W. T. (5). *Discussion*, p. 260.
[6] Suess (1).

favour of "Gondwana land," and compares this continent to "Atlantis" in the Northern Hemisphere. He lays stress upon the undeniable similarity in the structure of South Africa and the Indian Peninsula.

The idea of a connection between South Africa and India is supported by Stow, Blanford, Griesbach and others. Strong arguments, on the other hand, have been advanced by Wallace against the existence of "Lemuria," and these have been quoted as affording an argument against Gondwana land; but, as R. D. Oldham[1] suggests, the force of the argument against a land connection since Miocene times may be granted, but there is no reason to suppose that the present animals and plants can throw light upon the distribution of Secondary and Palæozoic land.

We have an able account of the Glossopteris flora and its lessons in W. T. Blanford's presidential address[2] to the Geological Section of the British Association meeting at Montreal. In that address he expresses his belief in a Southern continent, and considers we have strong evidence in favour of the existence of two distinct types of vegetation in Carboniferous times.

In his Presidential address before the Geological Society, Blanford[3] further dwells upon the evidence against a world-wide distribution of the European Coal-measures flora.

After speaking of the Glossopteris flora he remarks "Now this flora is so strongly contrasted with the Carboniferous flora of Europe that it is difficult to conceive that the countries in which the two grew can have been in connection, and the hypothesis of Gondwana land, as it is termed by Suess, ... seems more in accordance with facts than Mr Wallace's view that 'frequently evidence derived from such remote periods' is 'utterly inconclusive'."

R. D. Oldham[4] believes we have evidence of extensive ice-action, and is of opinion that the Glossopteris flora, which replaced the typical Carboniferous flora in the Southern Hemisphere, was probably driven towards the Equator by the low temperature.

On the other hand it may be, as Neumayr[5], Blanford[6] and Feistmantel have clearly indicated, that we have to connect the

---

[1] Oldham (2).
[2] Blanford, W. T. (4).
[3] Blanford, W. T. (7), p. 96.
[4] Oldham (2).
[5] Neumayr (2), (3).
[6] Blanford, W. T. (7), p. 96.

appearance of the Glossopteris flora with the widespread boulder beds, and regard the development of this new type of vegetation as the result of the cold climate. This new flora shewed itself vigorous in replacing the older Carboniferous types and gradually spread towards the North.

Neumayr considers the vigour of the Glossopteris flora suggests continental rather than insular vegetation, and adds this as an additional argument for the existence of a great Southern continent.

CHAPTER VIII.

PLEISTOCENE PLANTS, AND CONCLUSION.

A CONSIDERABLE amount of information upon Prehistoric floras is brought together by Mr Clement Reid[1] in a paper in the *Annals of Botany* for 1888. Plants are recorded from the Cromer beds of Norfolk and from other deposits of different ages.

The different deposits may be roughly divided into Postglacial, Interglacial, and Preglacial.

The Cromer Forest bed (Upper Pliocene) has been fully described in Mr Reid's able Memoir on *The Geology of the country around Cromer*[2].

In the flora of this Preglacial deposit species occur which are no longer found in Britain, and which point to very Arctic conditions previous to the formation of the first boulder clay.

Of interglacial age we have the plant-bearing beds near Edinburgh. All the plants from the Redhall quarries (3 miles from Edinburgh) are still native in the Scotch lowlands except Galeopsis tetrahit and Carum carui: the flora of this age as a whole suggests a climate somewhat colder than that of the South of Scotland at the present day.

From the submerged forest and other beds of Postglacial age have been obtained a number of species still living in Britain.

A conveniently arranged list is given of the plants mentioned in Reid's paper by " H. B. W." in the *Geological Magazine* for 1888, p. 567.

---

[1] Reid (2).   [2] Reid (1).

Since the publication of Clement Reid's summary in the *Annals of Botany* further additions have been made to our knowledge of Pleistocene vegetation in Britain. A number of plants are recorded by Reid and Ridley[1] from Hoxne, in Suffolk: the flora proved to be an Arctic one and is compared by the authors to the Iceland flora: the evidence is considered to point to "the approach of a warmer period following an Arctic one."

From other localities in Suffolk plants have recently been described[2] which differ from the Hoxne species in the absence of such Arctic forms as Betula Nana, Salix polaris, &c. The flora as a whole suggests a much less rigorous climate than that under which the leaf-bearing beds of Hoxne were deposited.

Plants suggestive of cold climates have been described by other writers from various countries, but these cannot be dealt with in the present sketch[3].

In an address delivered at the Birmingham meeting of the British Association Mr Carruthers[4] has given a general account of Pleistocene plants. In addition to the Cromer beds some lacustrine deposits at Holderness have yielded Arctic plants. Arctic or Northern plants are recorded too from Sweden; and Dawson[5] describes well-preserved northern species from the deposit of Leda clay at Green's Creek on the Ottawa River[6].

From this short sketch of Pliocene and post-Pliocene floras we see that Palæobotany supplies us with facts suggestive in some cases of colder conditions than the present, although the evidence afforded by fossil plants in the earlier geological periods is generally considered to point to temperatures in past eras higher than those of the same latitudes to-day.

Even in the more recent plant-bearing beds we have recently been reminded there is need of much caution in attempting to use plants as thermometers of climatal conditions. The Dürnten Lignites have yielded a number of plants, and these were considered by Heer to point to an "interglacial" climate similar

[1] Reid and Ridley (5), list of species given, pp. 443, 444.
[2] Candler (1), list of species, p. 507.
[3] Nathorst (2), Blytt (1).
[4] Carruthers (2), pp. 5—7.
[5] Dawson (2), p. 227.
[6] Since this was written additional facts in connection with Pleistocene floras have been recorded, *e. g.* Nathorst (3), Reid (4).

to that of Switzerland at the present day. Prof. Prestwich[1] has thrown doubt upon the interpretations of Heer, and considers them unwarranted by the facts upon which he relied: "Admitting the fact," says Prestwich, "that the lignite rests on beds of undoubted glacial (ground-moraine) origin, and that the trees grew on the spot where their stumps and remains are found, it by no means follows, as contended, that because these trees are all of species now living in Switzerland, the temperature was that of Switzerland of the present day. Pinus Sylvestris, Abies excelsa, the Yew, the Birch, and the Oak flourish equally in Sweden and far North in Siberia. On the other hand, there is one species of Pinus (P. montana) which is spread over the mountain country up to heights of 7000 feet, and is rare in the low lands; while one of the mosses is closely allied to a species now growing on the hills of Lapland.......Is the return, therefore, of the retreating glacier, supposing the boulder-gravel above the lignites of Dürnten to be due to direct ice-action, to be ascribed to anything more than a comparatively slight temporary change of climate, like those that now for a succession of seasons cause, from time to time, a temporary advance of the glaciers, only more marked?"

Conclusion.

No attempt has been made to discuss each geologic system in detail with a view to determine climatal conditions. The Carboniferous flora, or rather Permo-Carboniferous flora, has been treated at some length; as well as the several Arctic floras. In the case of the Cretaceous and Tertiary floras of Arctic regions, and indeed Tertiary floras generally, the conclusions as to climatal conditions are based upon safer evidence, and have therefore greater claim to be seriously considered, than the inferences drawn from older floras. But even in dealing with Tertiary plants it is difficult to separate those characters which are the result of climatal conditions, and those which should be regarded as expressions of the stage reached in the development or evolution of the floras. For example, in the Tertiary floras of Australia[2] and New Zealand[3] Ettingshausen has shewn that we have together a large number

[1] Bulman (1).　　[2] Ettingshausen (3).　　[3] Ettingshausen (4).

of plant types congregated together such as now characterise widely separated latitudes. Such a commingling of types, difficult to understand from the point of view of climate or other physical conditions of environment, must be explained by regarding such a synthetic type of flora as one not yet differentiated into its various branches.

In the case of the Triassic and Jurassic floras no detailed examination has so far been attempted, with a view to find out how far we may arrive at an estimate of the climate which obtained during those periods, and how far the climate was dependent upon geographical position.

A close examination of the Mesozoic floras, carried out in lines similar to those followed by Neumayr, in dealing with animal fossils, may possibly place us in a better position to discuss climatal conditions of Triassic, Jurassic and Cretaceous times. Such a method necessitates a thorough review of the several floras of the Mesozoic era. Work, at present in hand, may, however, supply the necessary data for such an attempt and enable us to discover how far it is possible to draw trustworthy conclusions from Mesozoic plant distribution as to the existence of climatal zones or botanical provinces.

In conclusion, we may briefly consider *Ferns*, as an example of a special class of plants, spoken of by many palæobotanists as trustworthy guides in climatal questions.

Sir Joseph Hooker[1] expressed the opinion that Ferns are less suitable as thermometers of past eras than other plants: but, as Baker[2] points out in his paper on the *Geographical Distribution of Ferns*, their distribution at the present day is well-known; this fact, coupled with their abundance as fossils, points to them as plants worthy of careful study for the purposes of our present enquiry.

Göppert[3] attempted in his *Systema Filicum Fossilium* to collect information upon fossil Ferns from all geological horizons, and gave a number of tables shewing the geographical range of the several species. Such a method of enquiry, granting the data reliable, should lead to useful results in considerations of past climates.

---

[1] Hooker (1).   [2] Baker (1), p. 305.   [3] Göppert (1).

Since Göppert's time our knowledge of fossil Ferns has made a certain advance, but we are still in the dark with regard to the real significance of many common fossil forms. This remark applies to Palæozoic, Mesozoic and Cainozoic Ferns.

The microscopical method of examination has established the fact that the Marattiaceæ were predominant in Coal-measure times.

Stur[1] gives a table which clearly demonstrates this: in his recent communication on Palæozoic Ferns this table is quoted by Bower[2]. It should be remembered, however, that such a table can only be regarded as affording a guess as to the relative proportions in which the several families were represented: in the Polypodiaceæ, for example, certain genera are included by Stur which, it is by no means certain, find their proper place there.

We may take it as an established fact that in the Carboniferous forests the Marattiaceæ occupied a prominent position. To-day this family is represented by the following genera:— Angiopteris, Marattia, Danæa and Kaulfussia; 21 species in all. Baker[3] gives their distribution as follows:—Frigid Zone, 0: North Temperate Zone, 2: South Temperate Zone, 3: Torrid Zone, 22.

Does the fact that the Marattiaceæ, now characteristic of the Torrid Zone, were the predominant family in Coal-measure times, prove a higher temperature in these latitudes during the Carboniferous period than at the present day?

Grand'Eury[4] recognises the possibility that the genera representing the family in the Palæozoic era may have been able to live under different conditions than those most favourable to their living descendants. We must not conclude from such facts of distribution as are afforded by this family of Ferns, fossil and recent, that the climate has changed in Europe from tropical to temperate. The present survivors of this once widespread and vigorous family cannot afford any strong argument as to the conditions of climate under which their more powerful ancestors were able to live.

If the material were to hand we might review the distribution of a family of Ferns—say the Marattiaceæ—in the Coal-measures,

[1] Stur (1), p. 411.   [2] Bower (1), p. 127.
[3] Baker (1), p. 351.   [4] Grand'Eury (1).

and notice how far latitude makes itself apparent in the proportion in which the family is represented in different parts of the world. From such a method of enquiry valuable results might be expected.

When we come to the question of the position occupied by Ferns, as a whole, in the Coal-measure vegetation, and compare it with their present position, we find a striking contrast, but in a period so far back as the Carboniferous no deductions of any great value in climatal speculations can be drawn from such comparisons. We have to take into account the facies of the whole flora and cannot draw any reliable conclusions from the consideration of one member of the flora, without having before us the vegetation *en bloc* and thus getting some idea of the surrounding conditions, as regards competing forms and the like, under which the members of the flora existed.

The *Coniferæ* have been studied more minutely than many families—both as regards fossils and living forms—and these may eventually lead us to valuable climatal retrospects.

There is the same confusion in nomenclature and synonyms in fossil Conifers as in other groups of fossil plants. The fact that cones, leaves and branches are nearly always found detached, has naturally led to a confusing multiplicity of terms. In cases where the internal structure is preserved, and in Conifers this is by no means rare, we may often determine with considerable accuracy the genus and species. In the more recent formations we may come within a measurable distance of reliable conclusions as to climate with these plants as our guides: this is illustrated in the case of the Amber beds whose Conifers have been studied in detail[1]. Fontaine[2] notices the fact that the North Carolina strata are richer in Conifers than those of Virginia. It is pointed out, and this may with advantage be borne in mind in dealing with fossil floras as test of climate, that this difference in the number of Conifers is probably only an accident of preservation. The conditions of sedimentation have much to do with the characters of fossil floras.

The geological histories of Sequoia[3] and Salisburia[4], so well

[1] Conwentz (1), Göppert and Menge (3).   [2] Fontaine (1), p. 124.
[3] Asa Gray (1).
[4] Heer. *Engler Bot. Jahrb.* 1881, p. 1.

told by Asa Gray and Heer, bring out in a striking manner the difference between the past and present geographical ranges of these genera.

We cannot, however, assert that in those localities where fragments of Salisburia and Sequoia are found embedded in the rocks climatal conditions obtained identical with, or even similar to, those which now obtain within the narrow limits of the homes of these two almost extinct genera.

In certain cases we may expect to gain information as to climates of the past, and the existence of botanical provinces, by taking each geological system by itself, and noting in its floras the geographical range of the several members. Our knowledge of the histology of fossil plants, especially those of Carboniferous age, has in recent years made very striking progress. Many botanists have at length recognised the importance of Palæobotany, and fossil plants have ceased to be regarded simply as aids, of no great value, to the stratigraphical geologist. We may expect, therefore, that a closer study of the geological floras, not only from phylogenetic and anatomical but also from biological points of view, may enable us to penetrate further into the life-conditions of those forests of which the earth's crust affords us such numerous though often too fragmentary relics.

LIST OF WORKS REFERRED TO IN THE TEXT.

Areschoug, F. W. C. (1) Der Einfluss des Klimas auf die Organisation der Pflanzen insbesondere auf die anatomische Structur der Blattorgane. *Engler. Bot. Jahrb. Leipzig.* Vol. II. 1882, p. 511.

Artis, F. T. (1) Antediluvian Phytology. *London*, 1825.

Asa Gray. See **Gray, Asa.**

Austen. See **Godwin-Austen.**

Baker, J. G. (1) On the geographical distribution of Ferns. *Trans. Linn. Soc. London*, 1870. Vol. XXVI. p. 305.

Balfour, J. H. (1) Palæontological Botany. *Edinburgh*, 1872.

Ball, J. (1) On the origin of the Flora of the European Alps. *Proc. R. Geogr. Soc. London*, 1879.

Bary, de. (1) Comparative Anatomy of the vegetative organs of the Phanerogams and Ferns. *Oxford*, 1884.

Bentham, G. (1) Presidential address before the Linnæan Society of London, 1871.

Binney, E. W. (1) Observations on the structure of fossil plants found in the Carboniferous strata. Part I. Calamites and Calamodendron. *Pal. Soc. London*, 1868.

Blanford, H. F. (1) On the age and correlation of the Plant-bearing series of India and the former existence of an Indo-Oceanic Continent. *Quart. Journ. Geol. Soc.* Vol. XXXI. 1875, p. 519.

Blanford, H. F. & **Medlicott, H. B.** (2) A manual of the Geology of India, *Calcutta*, 1879.

Blanford, W. T. (1) Note on the geological age of certain groups comprised in the Gondwana series of India, and on the evidence they afford of distinct Zoological and Botanical Terrestrial Regions in ancient epochs. *Rec. Geol. Surv. Ind. Calcutta.* Vol. IX. Pt. iii. p. 79, 1876.

────── (2) Palæontological relations of the Gondwana System : a reply to Dr Feistmantel. *Rec. Geol. Surv. Ind.* Vol. XI. Pt. i. p. 104, 1878.

────── (3) Notes on a character of the Talchir boulder beds. *Rec. Geol. Surv. Ind.* Vol. XX. Pt. i. p. 49, 1887.

────── (4) Address to the Geological Section of the British Association. *Montreal*, 1884.

────── (5) On additional evidence of the occurrence of Glacial Conditions in the Palæozoic Era, and on the Geological Age of the Beds containing

Plants of Mesozoic type in India and Australia. *Quart. Journ. Geol. Soc. London*, 1886. Vol. XLII. p. 249.

Blanford, W. T. & Blanford, H. F. & Theobald, W. jun. (6) On the geological structure and relations of the Talcheer Coal Field, in the district of Cuttack. *Mem. Geol. Surv. India. Calcutta*, 1859. Vol. I. p. 33.

Blanford, W. T. (7) Presidential address before the Geological Society of London. 1890.

Blytt, A. (1) On variations of climate in the course of time. *Christiania Videnskabs-Selskabs Forhandlinger*, 1886, No. 8. *Christiania*.

Boué, A. (1) On the changes which appear to have taken place during the different periods of the earth's formation, in the climate of our globe, and in the nature, and the physical and geographical distribution of its animals and plants. *Edinburgh Phil. Journ.* Vol. I. p. 88, 1826.

Bower, F. O. (1) Is the Eusporangiate or the Leptosporangiate the more primitive type in the Ferns? *Annals of Bot. Oxford.* Vol. V. p. 109, 1891.

Brongniart, Adolphe. (1) Considérations générales sur la nature de la végétation qui couvrait la surface de la terre aux diverses époques de formation de son écorce. *Ann. Sci. Nat. Paris.* Vol. XV. 1828. (Translated in *Edinburgh Phil. Journ.* 1828, vi. p. 349.)

―――― (2) Prodrome d'une Histoire des végétaux fossiles. *Paris*, 1828.

―――― (3) Histoire des végétaux fossiles. *Paris*, 1828―37.

Brongniart, Alex. (1) On fossil vegetables traversing the beds of the Coal-measures. *Ann. Mines. Paris*, 1821. (Translated by De la Beche in *Selections of Geological Memoirs from the Ann. Mines. London*, 1824.)

Buchanan, F. (1) A uniformity of climate prevailed over the Earth prior to the time of the Deluge? *Edinburgh Phil. Journ.* 1829, 8, p. 366.

Bulman, G. W. (1) On the sands and gravels in the boulder clay. *Geol. Mag. London*, 1891. p. 337.

Bunbury, C. J. F. (1) Botanical fragments. *London*, 1883.

Candler, C. (1) Observations on some undescribed lacustrine deposits at Saint Cross, South Elmham, in Suffolk. *Quart. Journ. Geol. Soc.* Vol. XLV. p. 504, 1889.

Candolle, A. de. (1) Géographie botanique raisonnée, ou exposition des faits principaux et des lois concernant la distribution géographique des plantes de l'époque actuelle. *Paris*, 1855.

Carruthers, W. (1) On the history, histological structure and affinities of Nematophycus Logani (Carr.) (Prototaxites Logani [Dawson]), an Alga of Devonian age. *Monthly Mic. Journ. London.* Vol. VIII. p. 160, 1872.

―――― (2) Address to the Biological Section of the British Association. *Birmingham*, 1886.

―――― (3) The Cryptogamic Forests of the Coal Period. *Geol. Mag.* 1869, Vol. VI. p. 289.

―――― (4) On Caulopteris punctata (Goepp.), a tree-fern from the Upper Greensand of Shaftesbury in Dorsetshire. *Geol. Mag.* 1865, p. 484.

Caspari, H. (1) Beiträge zur Kenntniss des Haut-Gewebes der Cacteen. *Halle,* 1883 (*Diss.*). *Bot. Zeitung. Berlin,* 1885. p. 804.

Clarke, W. B. (1) On the relative positions of certain plants in the Coal-bearing beds of Australia. *Quart. Journ. Geol. Soc.,* Vol. XVII. 1861, p. 354.

———— (2) On the Coal seams near Stony Creek, N. S. Wales. *Trans. R. Soc. Victoria. Melbourne,* 1861—64. p. 27.

Conwentz, H. (1) Monographie der Baltischen Bernsteinbäume. *Danzig,* 1890.

Corda, A. J. (1) Beiträge zur Flora der Vorwelt. *Prag,* 1845.

Credner, H. (1) Elemente der Geologie. *Leipzig,* 1887.

Crichton, Alex. (1) On the climate of the antediluvian world, and its independence of solar influence, and on the formation of Granite. *Annals of Philosophy. London,* 1825. p. 97.

Croll, J. (1) Discussions on Climate and Cosmology. *Edinburgh,* 1885.

———— (2) On the physical cause of the change of climate during Geological Epochs. *Phil. Mag. London.* Vol. XXVIII. p. 121, 1864.

Dana, J. D. (1) Manual of Geology. *Philadelphia,* 1863.

David, T. W. E. (1) Evidence of glacial action in the Carboniferous and Hawkesbury series, N. S. Wales. *Quart. Journ. Geol. Soc.,* Vol. XLIII. p. 190, 1887.

Dawes, J. S. (1) Further remarks upon the Calamite. *Quart. Journ. Geol. Soc.,* Vol. VII. p. 196, 1851.

Dawson, J. W. (1) The fossil plants of the Devonian and Upper Silurian formations of Canada. *Rept. Geol. Surv. Canada (Montreal),* 1871.

———— (2) The Geological history of plants. *London,* 1888.

Drude, O. (1) Handbuch der Pflanzengeographie. *Stuttgart,* 1890.

———— (2) Betrachtungen über die hypothetischen vegetationslosen Einöden im temperierten Klima der nördlichen Hemisphäre zur Eiszeit. *Petermann: Geograph. Mittlgn. Gotha,* 1889. Vol. XII. p. 282.

Dunn, E. J. (1) Notes on the occurrence of glaciated pebbles in the so-called Mesozoic conglomerate of Victoria. *Trans. R. Soc. Victoria. Melbourne.* Vol. XXIV. i. p. 44, 1887.

Dyer, W. T. Thiselton. (1) Article on Distribution (plants). *Encyclopædia Britannica,* Ed. 9, Vol. VII. 1877. *Edinburgh.*

Elliot, G. F. Scott. See **Scott-Elliot.**

Engler, A. (1) Versuch einer Entwicklungsgeschichte der Pflanzenwelt, insbesondere der Florengebiete seit der Tertiärperiode. *Leipzig,* 1879—82.

Ettingshausen, C. von. (1) Die Farnkräuter der Jetzwelt. *Vienna,* 1865.

———— (2) Die Blattskelete der Dikotyledonen. *Vienna,* 1861.

———— (3) Die genetische Gliederung der Flora Australiens. *Denksch. k. Ak. Wiss. Math.-Natur. Cl. Vienna,* 1874.

———— (4) Beiträge zur Kenntniss der fossilen Flora Neuseelands. *Denksch. k. Ak. Wiss. Math.-Natur. Cl. Vienna,* 1887.

Fedden, F. (1) On the evidences of 'ground-ice' in tropical India during the Talchir period. *Rec. Geol. Surv. India.* Vol. VIII. Pt. i. p. 16, 1875.

Feilden, H. W. & De Rance, C. E. (1) Geology of the coasts of the Arctic lands visited by the late British Expedition under Capt. Sir Geo. Nares, R.N. &c. *Quart. Journ. Geol. Soc.*, Vol. XXXIV. p. 556, 1878.

Feistmantel, O. (1) Notes on the age of some fossil floras in India. *Rec. Geol. Surv. Ind.*, Vol. IX. Pt. ii. p. 28 and Pt. iii. p. 63, 1876.

—— (2) Notes on the age of some fossil floras in India. *Rec. Geol. Surv. Ind.*, Vol. IX. Pt. iv. 1876, p. 115.

—— (3) On the Gondwana series of India as a probable representative of the Juro-Triassic epoch in Europe. *Geol. Mag.* 1876, p. 481.

—— (4) Contributions towards the knowledge of the fossil floras in India. *Journ. As. Soc. Bengal. Calcutta.* Vol. XLV. Pt. ii. p. 329, 1876.—A sketch of the history of the fossils of the Indian Gondwana system. *Journ. As. Soc. Bengal.* Vol. L. Pt. ii. p. 168, 1881.

—— (5) Ueber die pflanzen- und kohlenführenden Schichten in Indien, Afrika und Australien, und darin vorkommende glaciale Erscheinungen. *Sitz. k. böhm. Ges. Wiss. Prague,* 1887. p. 3.

—— (6) Geologische und palæontologische Verhältnisse der kohlen- und pflanzenführenden Schichten im östlichen Australien. *Sitz. k. böhm. Ges. Wiss. Prague,* 1887. p. 717.

—— (7) Ueber die geologischen und palæontologischen Verhältnisse des Gondwana-system in Tasmanien und Vergleichung mit andern Ländern, nebst einem systematischen Verzeichniss der im Australischen Gondwana-system vorkommenden Arten. *Sitz. k. böhm. Ges. Wiss.* 1888. p. 584.

Felix, J. (1) Untersuchungen über den inneren Bau westfälischer Carbon-Pflanzen. *Jahrb. k. preuss. geol. Landesanst. Berlin,* 1886, p. 1.

Fischer, H. (1) Ein Beitrag zur vergleichenden Anatomie des Markstrahlgewebes und der jährlich Zuwachszonen im Holzkörper von Stamm, Wurzel und Aeste bei Pinus Abies. *Flora, Regensburg,* 1885, p. 302.

Fleming, J. (1) On the value of the evidence from the Animal Kingdom tending to prove that the Arctic regions formerly enjoyed a milder climate than at present. *Edinburgh, Phil. Journ.* 1828, vi. p. 277.

Fontaine, W. M. (1) Contributions to the knowledge of the older Mesozoic flora of Virginia. *Monographs. U. S. Geol. Surv.* Vol. VI. 1883. *Washington.*

—— (2) The Potomac or younger Mesozoic flora. *Monographs. U. S. Geol. Surv.* Vol. XV. 1889.

Forbes, E. (1) The geological relations of the existing fauna and flora of the British Isles. *Mem. Geol. Surv. Great Britain.* Vol. I. p. 336, 1846.

Fuchs, T. (1) Die Mediterranflora in ihrer Abhängigkeit von der Bodenunterlage. *Sitz. k. Ak. Wiss. Vienna. Math.-Nat. Cl.* Vol. LXXVI. Abth. I. p. 240, 1877.

Gardner, J. Starkie. (1)   Are the fossil floras of the Arctic regions Eocene or Miocene? *Nature. London.* Vol. XIX. p. 124, 1878.

—— (2) On the Leaf-beds and Gravels of Ardtun, Carsaig, &c. in Mull. *Quart. Journ. Geol. Soc.* Vol. XLIII. p. 270, 1887.

—— (3) Remarks on some fossil leaves from the island of Mull. *Journ. Linn. Soc. London.* Vol. XXII. p. 219, 1887.

—— (4) The tropical forests of Hampshire. *Geol. Mag.*, 1877, p. 23.

Geikie, A. (1) Text-book of Geology. *London*, 1882.

Geikie, J. (1) On the buried Forests and Peat mosses of Scotland. *Trans. R. Soc. Edinburgh.* Vol. XXIV. p. 363, 1867.

—— (2) Evolution of climate. *Scottish Geog. Mag. Edinburgh.* Vol. VI. p. 57, 1890.

Gilpin, T. (1) An Essay on organic remains, as connected with an ancient tropical region of the Earth. *Philadelphia*, 1843.

Godwin-Austen, R. A. (1) On the valley of the English Channel. *Quart. Journ. Geol. Soc.* Vol. VI. p. 69, 1850.

—— (2) Extension of the Coal-measures beneath the South-Eastern part of England. *Quart. Journ. Geol. Soc.* Vol. XII. p. 38, 1856.

Goebel, K. (1) Pflanzenbiologische Schilderungen. Pts. i. and ii., 1889— 91. *Marburg.*

Göppert, H. R. (1) Systema filicum fossilium. *Nova Acta Ac. Cœs. Leop.- Car. Halle.* Vol. XVII. Supplement, 1836.

—— (2) Ueber das Gefrieren, Erfrieren der Pflanzen und Schutzmittel dagegen. *Stuttgart*, 1883.

Göppert, H. R., & Menge, A. (3) Die Flora des Bernsteins und ihre Beziehungen zur Flora der Tertiärformation und der Gegenwart. Vol. I. 1883, vol. II. 1886. *Danzig.*

Grand'Eury, C. (1) Flore Carbonifère du Département de la Loire et du centre de la France. *Mém. Ac. Sci. Paris.* Vol. XXIV.

Gray, Asa. (1) Sequoia and its history. Collected papers : scientific papers of Asa Gray, selected by C. S. Sargent. *London*, 1889. Vol. II. p. 142.

—— (2) Plant Archæology. (Review of Saporta's Le Monde des plantes avant l'apparition de l'homme.) Collected papers. Vol. I. p. 269.

Greely, A. W. (1) Three years of Arctic Service. *London*, 1886.

Green, A. H. (1) A contribution to the Geology and Physical Geography of the Cape Colony. *Quart. Journ. Geol. Soc.*, Vol. XLIV. p. 239, 1888.

Griesbach, C. L. (1) On the geology of Natal, in South Africa. *Quart. Journ. Geol. Soc.*, Vol. XXVII. p. 58, 1871.

—— (2) Geological notes. *Rec. Geol. Surv. Ind.*, Vol. XIII. Pt. ii. p. 83, 1880.

—— (3) Afghan and Persian field-notes. *Rec. Geol. Surv. Ind.*, Vol. XIX. Pt. i. p. 48, 1886.

—— (4) Field notes to accompany a geological sketch map of Afghanistan and Khorassan. *Rec. Geol. Surv. Ind.*, Vol. XX. Pt. ii. p. 93, 1887.

Grisebach, A. (1) Die Vegetation der Erde nach ihrer klimatischen Anordnung. *Leipzig*, 1884.

Haberlandt, G. (1) Vergleichende Anatomie des assimilatorischen Gewebesystems der Pflanzen. *Pringsheim. Jahrb. Berlin*, Vol. XIII. p. 74, 1882.

—— (2) Die Physiologischen Leistungen der pflanzengewebe. *Handbuch der Botanik. Schenk. Breslau*, 1882.

Haughton, S. (1) Physical geology. *Nature*, 1878, p. 266.

Heer, O. (1) The Primæval world of Switzerland. Edit. by J. Heywood, *London*, 1876.

—— (2) Flora tertiaria Helvetiæ. *Winterthur*, 1855—59.

—— (3) Flora fossilis Arctica : die fossile Flora der Polarlände. *Zürich*, 1868—83.

Hennessy, H. (1) On Terrestrial climate as influenced by the distribution of land and water during different geological periods. *Atlantis*, Vol. II. p. 208. *London*, 1859.

Hildebrand, F. R. (1) Die Lebensdauer und Vegetationsweise der Pflanzen, ihre Ursachen und ihre Entwickelung. *Engler, Bot. Jahrb.*, Vol. II. p. 51, 1882.

Hochstetter, F. von. (1) New Zealand. *Stuttgart*, 1867.

Hoffmann, F. (1) Bemerkungen über die gegenseitige Verhältnissen der vorweltlichen Floren. *Ann. Phys. Chem. Leipzig*. Vol. XV. p. 415, 1829.

Holmes, T. V. (1) Presidential Address before the Geologists' Association, 1891.

Hooker, J. D. (1) On the vegetation of the Carboniferous period as compared with that of the present. *Mem. Geol. Surv. Great Britain.* Vol. II. Pt. ii., 1848.

—— (2) The Botany of the Antarctic voyage of H.M. Discovery ships, Erebus and Terror, in the years 1839—43. Flora Antarctica. *London*, 1847.

Hunt, T. Sterry. (1) Geological and Chemical Essays. *Salem*, 1878.

Huxley, T. H. (1) Presidential Address before the Geological Society, 1862.

Johnston, H. H. The Kilima-njaro Expedition. *London*, 1886.

Johnston, R. M. (1) General observations regarding the classification of the upper Palæozoic and Mesozoic rocks of Tasmania, together with a full description of all the known Tasmanian Coal plants, including a considerable number of new species. *Proc. R. Soc. Tasmania. Hobart Town*, 1885, p. 343.

Jones, Rupert. See **Rupert Jones.**

Judd, J. W. (1) Presidential Address before the Geological Society, 1888.

Jussieu, A. de. (1) Examen des causes des impressions des plantes marquées sur certaines pierres des environs de Saint-Chaumont dans le Lyonnais. *Mém. Ac. Sci. Paris*, 1718, p. 287.

Karr, H. W. Seton. See **Seton-Karr.**

Kerner von Marilaun, A. (1) Studien über die Flora der Diluvialzeit in

140   LIST OF WORKS REFERRED TO IN THE TEXT.

den östlichen Alpen. *Sitz. k. Ak. Wiss. Vienna. Math.-Nat. Cl.,* Vol. XCVII. 1889.

Kerner von Marilaun, A. (2) Pflanzenleben. *Leipzig,* 1888.

Kidston, R. (1) Catalogue of the Palæozoic plants in the department of Geology and Palæontology, British Museum. *London,* 1886.

Kihlman, A. O. (1) Pflanzenbiologische Studien aus russisch. Lappland. *Helsingfors,* 1890.

Kjellman, F. R. (1) Ueber die Algenvegetation des murmanschen Meeres an der Westküste von Nowaja Semlja und Wajgatsch. *Nova Acta R. Soc. Upsala,* 1877 (vol. extra). See also *Compt. Rend. Paris,* Vol. 80, 1875. Végétation hivernale des Algues à Mosselbay (Spitzberg), d'après les observations faites pendant l'expédition polaire Suédoise en 1872—73.

Knowlton, F. H. (1) The fossil wood and Lignites of the Potomac formation. *American Geologist. Minneapolis.* Vol. III. ii. p. 99, 1889.

—— (2) Fossil wood and Lignite of the Potomac formation. *Bull, U. S. Geol. Surv. Territories.* No. 56. *Washington,* 1889.

—— (3) A revision of the genus Araucarioxylon of Kraus, with compiled descriptions and partial synonomy of the species. *Proc. U. S. Nat. Mus.* Vol. XII. p. 601. *Washington,* 1890.

Kny, L. (1) Ueber die Verdoppelung des Jahresringes. *Verh. des Bot. Ver. Brandenburg.* Vol. XXI. 1880.

Koninck, de. (1) Notice of De Koninck's book on fossils (Palæozoic) of N.S. Wales. *Geol. Mag.,* 1876, p. 364.

Krabbe, G. (1) Ueber die Beziehungen der Rindenspannung zur Bildung der Jahrringe und zur Ablenkung der Markstrahlen. *Sitz. d. k. preuss. Akad. d. Wissens. Berlin,* 1882, p. 1093.

Krüger, P. (1) Die oberirdischen Vegetationsorgane der Orchideen in ihren Beziehungen zu Clima und Standort. *Flora. Regensburg,* 1883, p. 435.

Lapierre, M. E. (1) Note sur le bassin houiller de Tete (Zambèze). Plants by Prof. Zeiller. *Ann. Mines. Sér.* viii. IV. p. 594. *Paris,* 1883.

Lesquereux, L. (1) Descriptions of fossil plants in *The Geology of Pennsylvania.* H. D. Rogers. *A Government Survey. Philadelphia.* Vol. II. Pt. ii., 1858.

Lindley, J. (1) Observations upon the effects produced on plants by the frost which occurred in England in the winter of 1837—8. *Trans. Hort. Soc. London,* 2nd Ser. Vol. II. p. 225, 1842.

Lindley, J., & Hutton, W. (2) The fossil flora of Great Britain. *London.* 1831—37.

Ludwig, R. (1) Pflanzen aus dem Rothliegenden im Government Perm. *Palæontographica,* Vol. X. p. 270. *Cassel,* 1861.

Lydekker, R. (1) Note on the Gondwana Homotaxis. *Rec. Geol. Surv. Ind.* Vol. XIX. Pt. ii. p. 133, 1886.

Lyell, C. (1) Principles of Geology. *London,* 1867.

Lyell, C. (2) The Student's Elements of Geology. *London*, 1878.

Marion, A. F. See **Saporta**.

Marr, J. E. (1) The work of Ice-Sheets. *Geol. Mag.*, 1887, p. 1.

Meddelelser om Grønland. Meddelelser om Grøn. udgivne af Commissionen for Ledelsen af de geologiske og geographiske Undersøgelser i Grønland. *Copenhagen*, 1879— .

Medlicott, H. B. (1) Memorandum on the discussion regarding the boulder beds of the Salt range. *Rec. Geol. Surv. Ind.* Vol. xix. Pt. ii. p. 131, 1886.

Medlicott and Blanford. (2) See **Blanford, H. F.** (2)

Moulle, M. A. (1) Mém. sur la géologie générale et sur les mines de diamants de l'Afrique du Sud. *Ann. Mines.* Ser. viii. Vol. vii. 1885, p. 193.

Nansen, F. (1) First crossing of Greenland. *London*, 1890.

Nathorst, A. G. (1) Kritische Bemerkungen über die Geschichte der Vegetation Grönlands. *Engler. Bot. Jahrb.*, Vol. xiv. 1892, p. 183.

———— (2) Ueber neue Funde von fossilen Glacialpflanzen. *Engler. Bot. Jahrb.*, Vol. i. 1882, p. 431.

———— (3) Fresh evidence concerning the distribution of Arctic plants during the Glacial Epoch. *Nature*, Jan. 21, 1892, p. 273.

———— (4) Bemerkungen über Professor Dr O. Drude's Aufsatz, "Betrachtungen über die hypothetischen vegetationslose Einöden im temperierten Klima der nördlichen Hemisphäre zur Eiszeit." *Engler. Bot. Jahrb.*, 1891. Vol. xiii. *Beiblatt* 29, p. 53.

Naudin, C. (1) Observations météorologiques à Collioure. *Ann. Sci. Nat. Paris.* Ser. 6, 1878, p. 337.

Neumayr, M. (1) Ueber klimatische Zonen während der Jura- und Kreidezeit. *Denkschr. k. Ak. Wiss. Vienna. Math.-Nat. Cl.* Vol. xlvii. 1883.

———— (2) Erdgeschichte. *Leipzig*, 1887.

———— (3) The Climates of Past Ages. *Nature*, 1890, pp. 148, 175.

Newberry, J. S. (1) Rhætic plants from Honduras. *Amer. Journ. Newhaven.* Vol. xxxvi. 1888, p. 342.

Noack, F. (1) Der Einfluss des Klimas auf die Cuticularisation und Verholzung der Nadeln einiger Coniferen. *Pringsheim. Jahrb.* Vol. xviii. 1887, p. 519.

Nordenskiöld, A. E. (1) On the former climate of the Polar regions. *Geol. Mag.*, 1875, p. 525.

Oldham, R. D. (1) Memorandum on the correlation of the Indian and Australian Coal-bearing beds. *Rec. Geol. Surv. Ind.* Vol. xix. Pt. i. p. 39, 1886.

———— (2) Some rough notes for the construction of a Chapter in the History of the Earth. *Journ. As. Soc. Beng.* No. iii. 1884, p. 187.

Oldham, R. D. (3) A note on the Olive group of the Salt Range. *Rec. Geol. Surv. India.* Vol. XIV. Pt. ii. p. 127, 1886.

Parkinson, J. (1) Organic remains of a former World. *London*, 1804—11.

Penhallow, D. P. (1) On Nematophyton and allied forms from the Devonian (Erian) of Gaspé and Bay des Chaleurs. *Trans. R. Soc. Canada, Montreal.* Vol. VI. 1888, p. 27.

Pfeffer, W. (1) Pflanzenphysiologie. *Leipzig*, 1881.

Phillips, J. (1) Manual of Geology. Vol. I. Edited by Prof. H. G. Seeley. *London*, 1885.

Pick, H. (1) Ueber den Einfluss des Lichtes auf die Gestalt und Orientirung der Zellen der Assimilationsgewebe. *Bot. Centralblatt. Cassel*, 1882, p. 400.

Pictet, F. G. (1) Traité de Paléontologie. *Paris*, 1853—57.

Potonié, H. (1) Die fossile Pflanzen-Gattung Tylodendron. *Jahrb. k. preuss. geol. Landesanst. Berlin*, 1887, p. 311.

Reid, Clement. (1) The Geology of the country around Cromer. *Mem. Geol. Surv.*, 1882.

—— (2) Notes on the Geological history of the recent Flora of Britain. *Annals Bot.*, Vol. II. 1888—9, p. 177.

—— (3) The origin of the Flora of Greenland. *Nature*, Vol. XLIV. p. 299, 1891.

—— (4) The Climate of Europe during the Glacial Epoch. *Natural Science*, Vol. I. 1892, p. 427.

Reid, Clement, & Ridley, H. N. (5) Fossil Arctic plants from the lacustrine deposits at Hoxne, in Suffolk. *Geol. Mag.*, 1888, p. 441.

Renault, B. (1) Structure comparée de quelques tiges de la Flore carbonifère. *Nouv. Arch. Mus. Paris*, Sér. ii. Vol. II. p. 213, 1879.

—— (2) Cours de Botanique fossile. *Paris*, 1881—85.

Richthofen, F. von. (1) China. Vol. IV. Pflanzliche Versteinerungen. Schenk. *Berlin*, 1883.

Rothpletz, A. (1) Ueber die palæozoischen Landpflanzen und ihre Verbreitungsgebiete. *Botanisches Centralblatt*, Vols. XXIX.—XXX., p. 283.

—— (2) Oswald Heer (obituary notice). *Bot. Cent.* Vol. XVII. p. 157, 1884.

Rupert Jones. (1) Manual of the Natural History, Geology, and Physics of Greenland and the neighbouring regions. Prepared for the use of the Arctic Expedition of 1875. *London*, 1875.

Sachs, J. (1) Text-book of Botany. *Oxford*, 1882.

—— (2) Lectures on the Physiology of Plants. *Oxford*, 1887.

Sanio, C. (1) Ueber den Bau der Jahrringe. *Bot. Zeitung. Leipzig*, 1863, p. 391.

Saporta, G. de. (1) Le monde des plantes avant l'apparition de l'homme. *Paris*, 1879.

Saporta, G. de. (2) La végétation du Globe dans les temps avant à l'homme. *Rev. Deux Mondes. Paris*, 1868, p. 315.

—— (3) Révision de la Flore des Gypses d'Aix. (Études sur la végétation du sud-est de la France à l'époque tertiaire.) *Ann. Sci. Nat.*, 1872. Vols. XV.—XVI. p. 277.

—— (4) Dernières adjonctions à la flore fossile d'Aix-en-Provence. *Paris*, 1889.

Saporta, G. de, et Marion, A. F. (5) Recherches sur les végétaux fossiles de Meximieux. *Arch. Mus. d'hist. nat. Lyons*, 1875.

Saporta, G. de. (6) La formation de la houille. *Rev. Deux Mondes*, 1882, p. 5.

Schenck, A. (1) Die geologische Entwickelung Süd-Afrikas. *Pet. Mitthlg.*, 1888, p. 225.

—— (2) *Protokoll der März-Sitzung. N. Jahrb. Stuttgart*, 1889, p. 172.

Schenck, H. (1) Die Biologie der Wassergewächse. *Bonn*, 1886.

—— (2) Vergleichende Anatomie der submersen Gewächse. *Cassel*, 1886. *Bibliotheca Botanica. Abhandlungen aus dem Gesammtgebiete der Botanik. Uhlworm und Haenlein.* Heft I.

—— (3) Ueber das Aërenchyma, ein dem Kork homologes Gewebe bei Sumpfpflanzen. *Pringsheim. Jahrb.* Vol. XX. 1889, p. 526.

Schenk, A. (1) Beiträge zur Flora der Vorwelt. IV. Die Flora der nordwestdeutschen Wealdenformation. *Palæontographica*, Vol. XIX. 1874, p. 203.

—— (2) Die fossile Flora der Grenzschichten des Keupers und Lias Frankens. *Wiesbaden*, 1867.

—— (3) Handbuch der Paläontologie : A. Zittel. Abth. II. Palæophytologie. *Munich & Leipzig*, 1890.

—— (4) See **Richthofen.**

Schimper, A. F. W. (1) Ueber Schutzmittel des Laubes gegen Transpiration, besonders in der Flora Java's. *Sitz. d. Ak. zu Berlin*, 1890—92, p. 1045.

Schimper, W. P. (1) Traité de paléontologie végétale, ou la flore du monde primitif dans ses rapports avec les formations géologiques et la flore du monde actuel. *Paris*, 1869—74.

Schleiden, M. J. (1) Grundzüge der wissenschaftlichen Botanik. *Leipzig*, 1845.

Schlotheim, E. von. (1) Die Petrefactenkunde auf ihrem jetzigen Standpunkte durch die Beschreibung seiner Sammlung versteinerter und fossiler Ueberreste des Thier- und Pflanzenreichs der Vorwelt erläutert. *Gotha*, 1820—23.

Schwendener, S. (1) Das mechanische Princip im anatomischen Bau der Monocotylen mit vergleichenden Ausblicken auf die übrigen Pflanzenklassen. *Leipzig*, 1874.

Scott-Elliott. (1) The effect of exposure on the relative length and breadth of leaves. *Journ. Linn. Soc. London*, 1890—91, p. 375.

Seton-Karr, H. W. Shores and Alps of Alaska. *London*, 1887.

Seton-Karr, H. W. (2) The Alpine regions of Alaska. *Proc. R. Geogr. Soc.*, 1887. Vol. IX. p. 269.

Solms, H. Graf zu. (1) Die Coniferenformen des deutschen Kupferschiefer und Zechsteins. *Pal. Abh. von Dames und Kayser.* Vol. II. Heft II. *Berlin*, 1884.

———— (2) Fossil Botany. *Oxford*, 1891.

Sorauer, P. (1) Der Einfluss der Luftfeuchtigkeit. *Bot. Zeitung*, 1878, p. 1.

Stahl, E. (1) Ueber den Einfluss des sonnigen oder schattigen Standortes auf die Ausbildung der Laubblätter. *Jenaische Zeit. für Naturwissenschaft. Jena*, Vol. XVI. 1883, p. 162.

———— (2) Ueber den Einfluss der Lichtintensität auf Structur und Anordnung des Assimilations-Parenchym. *Bot. Zeitung*, 1880, p. 868.

Steinhauer, H. (1) On fossil reliquia of unknown vegetables in the Coal strata. *Trans. Amer. Phil. Soc. Philadelphia*, Vol. I. 1818, p. 265.

Stenzel, G. (1) Nachträge zur Kenntniss der Coniferenhölzer der Palæozoischen Formationen. *Abh. k. Ak. Wiss. Berlin*, 1887, p. 3.

Sternberg, O. Graf von. (1) Versuch einer geognostisch-botanischen Darstellung der Flora der Vorwelt. *Leipzig*, 1820—38.

Strasburger, E. (1) Ueber den Bau und die Verrichtungen der Leitungsbahnen in den Pflanzen. *Histologische Beiträge.* Heft III. *Jena*, 1891.

Stur, D. (1) Die Carbonflora der Schatzlaren Schichten. I. Farne. *Abh. k.-k. Geol. Reichs. Vienna.* Vol. II. Abth. i. 1885.

———— (2) Die Lunzer- (Lettenkohlen) Flora in den "Older Mesozoic beds of the Coal-field of Eastern Virginia." *Verh. k.-k. Geol. Reichs. Vienna.* No. 10. 1888, p. 1.

Suess, E. (1) Das Antlitz der Erde. *Prague & Leipzig*, 1885.

Sutherland, —. (1) Notes on an ancient boulder-clay of Natal. *Quart. Journ. Geol. Soc.*, Vol. XXVI. p. 514, 1870.

Szajnocha, L. (1) Ueber fossile Pflanzenreste aus Cacheuta in der Argentinischen Republik. *Sitz. k. Ak. Wiss. Math.-Nat. Cl. Vienna.* Vol. XCVII. Abth. i. 1888, p. 1.

Tieghem, Ph. van. (1) Sur le ferment butyrique (Bacillus Amylobacter) à l'époque de la houille. *Compt. Rend. Paris.* Vol. LXXXIX. 1879, p. 1102.

———— (2) Traité de Botanique. *Paris*, 1891.

Topham, H. W. (1) An expedition to Mount St Elias, Alaska. *Alpine Journal. London*, 1889. Vol. XIV. p. 345.

Treub, M. (1) Notice sur la nouvelle Flore de Krakatau. *Bot. Cent.* Vol. XXXV. p. 298, 1885.

Tschirch, A. (1) Ueber einige Beziehungen des anatomischen Baues der Assimilationsgewebe zu Klima und Standort, mit specieller Berücksichtigung des Spaltöffnungs-Apparates. *Linnæa. Berlin.* Vol. IX. p. 139. 1880—82.

———— (2) Beiträge zu der Anatomie und dem Einrollungsmechanismus einiger Grasblätter. *Pringsheim. Jahrb.*, 1882. Vol. XIII. p. 544.

Unger, F. (1) Botanische Beobachtungen. *Bot. Zeitung*, 1847, p. 265.

Volkens, G. (1) Zur Kenntniss der Beziehungen zwischen Standort und anatomischen Bau der Vegetationsorgane. *Jahrb. des k. bot. Gartens und des bot. Mus. zu Berlin*, III. 1844.

—— (2) Zur Flora der Aegyptisch-arabischen Wüste. *Sitz. d. k. preuss. Akad. Wiss. zu Berlin*, 1886, p. 63.

Vries, H. de. (1) Ueber den Einfluss des Rindendruckes auf den anatomischen Bau des Holzes. *Flora*, 1875. p. 97.

—— (2) Ueber den Einfluss des Drucks auf die Ausbildung des Herbstholzes. *Flora*, 1877. p. 241.

Waagen, W. (1) Note on some Palæozoic fossils recently collected by Dr H. Warth in the Olive Group of the Salt Range. *Rec. Geol. Surv. Ind.*, Vol. XIX. Pt. i. p. 22, 1886.

—— (2) The Carboniferous Glacial Period. *Rec. Geol. Surv. Ind.*, Vol. XXI. Pt. iii. p. 89, 1888.

—— (3) The Carboniferous Glacial Period. Further note: on a letter from Mr C. Derby, concerning traces of a Carboniferous glacial period in S. America. *Rec. Geol. Surv. Ind.*, Vol. XXII. Pt. ii. p. 69, 1889.

—— (4) Note on the Bivalves of the Olive Group, Salt Range. *Rec. Geol. Surv. Ind.*, Vol. XXIII. Pt. i. p. 38, 1890.

Wallace, A. R. (1) Island Life, or the phenomena and cause of insular faunas and floras. *London*, 1880.

Ward, L. (1) Sketch of Paleobotany. *U. S. Geol. Surv. Fifth Ann. Rep.*, 1883—4. *Washington*.

—— (2) The geographical distribution of fossil plants. *U. S. Geol. Surv. Eighth Ann. Rep.*, 1887—88.

Warming, E. (1) Geschichte der Flora Grönlands. Antikritische Bemerkungen zu A. G. Nathorst's Aufsatz. *Engler. Jahrb.* Vol. XIV. 1892, p. 462.

Warth, H. (1) On the identity of the Olive series in the East, with the speckled sandstone in the West of the Salt Range. *Rec. Geol. Surv. Ind.*, Vol. XX. Pt. ii. p. 117, 1887.

—— (2) A facetted pebble from the boulder bed (speckled sandstone) of Mount Chel in the Salt Range in the Punjab. *Rec. Geol. Surv. Ind.*, Vol. XXI. Pt. i. p. 34, 1888.

Whitney, J. D. (1) The climatic changes of late geological times. *Mem. Mus. Harvard Coll.*, Vol. VII. No. ii. 1882. *Cambridge, U. S. A.*

Wieler, A. (1) Beiträge zur Kenntniss der Jahresringbildung und des Dickenwachstums. *Pringsheim. Jahrb.*, 1887. Vol. XVIII. p. 70.

Williamson, W. C. (1) A Monograph on the morphology and histology of Stigmaria ficoides. *Pal. Soc. London*, 1887.

—— (2) On the organisation of the fossil plants of the Coal-measures. *Phil. Trans. R. Soc. London.* Pt. i. 1871, Pt. ii. 1872, Pt. v. 1874, Pt. viii. 1877, Pt. ix. 1878.

S. E.   10

146 LIST OF WORKS REFERRED TO IN THE TEXT.

Witham, W. (1) The internal structure of fossil vegetables found in the Carboniferous and Oolitic deposits of Great Britain. *Edinburgh*, 1833.

Woeikof, A. Von. (1) Gletscher und Eiszeiten in ihrem Verhältnisse zum Klima. *Zeit. Gesell. für Erdkunde zu Berlin,* Vol. XVI. 1881, p. 217.

Wood, Searles (Junr). (1) The climate controversy. *Geol. Mag.*, 1876, p. 385.

—— (2) The tropical forests of Hampshire. *Geol. Mag.*, 1877, pp. 95 and 187.

Wright, G. F. (3) The Ice Age in North America. *London*, 1890.

Wynne, A. B. (1) On a certain fossiliferous Pebble-band in the Olive group of the Eastern Salt Range, Punjab. *Quart. Journ. Geol. Soc.*, Vol. XLII. 1886, p. 341.

—— (2) The Trans-Indus Salt region in the Kohát District. *Mem. Geol. Surv. India. Calcutta.* Vol. XI. Art. ii. 1875.

—— On the Geology of the Salt Range in the Punjab. *Mem. Geol. Surv. Ind.*, Vol. XIV. 1878.

—— On the Trans-Indus extension of the Punjab Salt Range. *Mem. Geol. Surv. Ind.*, Vol. XVII. Pt. ii. 1880.

Zeiller, R. (1) Observations sur quelques cuticles fossiles. *Ann. Sci. Nat.*, 1882, p. 217.

—— (2) See **Lapierre.**

INDEX.

CAMBRIDGE: PRINTED BY C. J. CLAY, M.A. AND SONS, AT THE UNIVERSITY PRESS.

Printed in the United States
By Bookmasters